WHALES, DOLPHINS AND MAN

WHALES, DOLPHINS AND MAN

Jacqueline Nayman

Hamlyn
London·New York·Sydney·Toronto

ACKNOWLEDGEMENTS

Colour

Ardea Photographics 75; Bruce Coleman Ltd. – J. and S. Brownlie 70, Sven
Gillsater 78, M. P. Harris 19 (top), Gordon Williamson 99 (top), 99 (bottom),
bottom centre front jacket, Russ Kinne 107 (top), James Simon 23 (bottom);
Ben Cropp 19 (bottom), 22, 30 and bottom right front jacket; Jose Dupont 18,
103 (bottom) and top left front jacket, 111 and top right front jacket; Hirmer
Fotoarchiv, Munchen 26; Institute of Oceanographic Sciences 67 (bottom),
79 (bottom), bottom left front jacket; Jacana Agence De Presse 23 (top), 31,
66, 67 (top), 103 (top), 107 (bottom); Keystone Press Agency 106, 110; Photo
Aquatics 71 (top), 71 (bottom), 79 (top), 98 and back jacket; Whaling Museum,
New Bedford, Massachusetts 74 (top), 74 (bottom); Z.E.F.A. 27, 102.

Black and White

Barnaby's Picture Library 62, 72 (top), 72 (bottom), 81, 88–89, 108, 115;
Bavaria Verlag 28, 32, 38, 39, 41, 100, 113, 117; Bruce Coleman Ltd. – Jen
and Des Bartlett 24, 52, Russ Kinne 12 (bottom right), 14, Vincent Serventy
47, 76; Institute of Oceanographic Sciences 15, 16, 50, 112; Fort Worth
Zoological Park 123; Knudsens Fotosenter 86; Frank W. Lane – 93,
Marineland 53, 55, 56, 57, 58, 77, 118–119, 120, Alfred Saunders 94; The
Mansell Collection 34–35, 37, 90; Miami Seaquarium 105; National Maritime
Museum 64; Paul Popper 44–45, 104, 124; Peabody Museum of Salem 83;
Radio Times Hulton Picture Library 29, 61, 65, 68, 69, 84; Wildlife Photos,
Clevedon 92, 96 (top) 96 (bottom).

It should be noted that the cetacean portraits, Copyright © The Hamlyn
Publishing Group Limited, were drawn from dead specimens and a uniform
scale was not employed.

Published by
THE HAMLYN PUBLISHING GROUP LIMITED
London · New York · Sydney · Toronto
Astronaut House, Feltham, Middlesex, England
Copyright © The Hamlyn Publishing Group Limited 1973
Reprinted 1978

ISBN 0 600 39274 0

Printed by New Interlitho, Milan, Italy.

CONTENTS

INTRODUCTION

There Leviathan
Hugest of living creatures, on the deep
Stretch'd like a promontory sleeps or swims,
And seems a moving land . . .
Milton (Paradise Lost)

Sixty years ago Scott, on his last, ill-fated voyage to the Antarctic, found time to write in the diary which later was to be found and read with such poignant interest, the following passage:

'1911, Thurs. Jan. 5 – All hands were up at 5:00 this morning and at work at 6:00 . . . . I was a little late on the scene this morning and thereby witnessed a most extraordinary scene. Some 6 or 7 killer whales, old and young, were skirting the fast floe edge ahead of the ship; they seemed excited and dived rapidly, almost touching the floe. As we watched, they suddenly appeared astern, raising their snouts out of the water. I had heard weird stories of these beasts, but had never associated serious danger with them. Close to the water's edge lay the wire rope of the ship, and our 2 Esquimaux dogs were tethered to this. I did not think of connecting the movement of the whales with this fact, and seeing them so close I shouted to Ponting, who was standing abreast of the ship. He seized his camera and ran towards the floe edge to get a close picture of the beasts, which had momentarily disappeared. The next moment the whole floe under him and the dogs heaved up and split into fragments. One could hear the 'booming' noise as the whales rose under the ice and struck it with their backs. Whale after whale rose under the ice, setting it rocking fiercely; luckily Ponting kept his feet and was able to fly to security. By an extraordinary chance also, the splits had been made around and between the dogs, so that neither of them fell into the water. Thus it was clear that the whales shared our astonishment, for one after another, their huge hideous heads shot vertically into the air through the cracks which they had made. As they reared them to a height of 6 or 8 feet it was possible to see their tawny head markings, their small glistening eyes and their terrible array of teeth – by far the largest and most terrifying in the world. There cannot be a doubt that they looked up to see what had happened to Ponting and

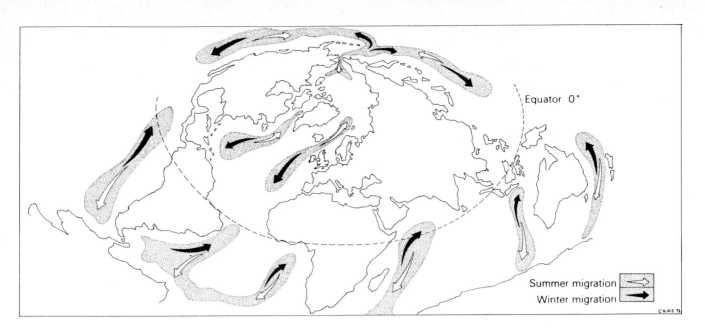

the dogs . . . . After this, whether they thought the game insignificant or whether they missed Ponting is uncertain, but the terrifying creatures passed on to other hunting grounds, and we were able to rescue the dogs . . . .

Of course, we have known well that killer whales continually skirt the edge of the floes and that they would undoubtedly snap up anyone who was unfortunate enough to fall into the water; but the fact that they could display such deliberate cunning, that they were able to break ice of such thickness (at least two and a half feet) and that they could act in unison, were a revelation to us. It is clear that they are endowed with singular intelligence and in the future we shall treat that intelligence with every respect.'★

Yet today killer whales are ridden by their trainers in many dolphin shows with apparently less danger than threatens the cowboy riding a bronco in a rodeo. The climax of the San Diego show comes when the trainer puts his head into the open mouth with its 'terrible array of teeth' and withdraws it safely to the 'Ooh!'s and 'Ah!'s of the crowd.

How do we reconcile these two accounts? Has the killer whale changed his character or is it man's attitude towards him that has changed?

Certainly there is a new consciousness of and sympathy towards whales of all sorts, particularly towards dolphins, which are close cousins to the killer whale *(Orcinus orca)* and are in fact themselves small whales. Thousands of people have been charmed by them in the dolphinariums and oceanariums which have recently opened in many countries. They have been charmed by their agility and singular intelligence and also by some sort of affinity which seems to exist between man and dolphin.

The ancient Greeks and Romans were well aware of this affinity and wrote of friendships between men and dolphins and incidents of dolphins helping drowning and drowned men to shore. These stories later came into disrepute. At the beginning of the nineteenth century a text-book of the day stated:

★Scott R. F. *Scott's Last Expedition*. John Murray. London. 1923

Conjectural migratory routes for rorquals and humpback whales. The humpback whale migrates nearer to land than do rorquals and its migratory routes are therefore better known. The northern and southern races of the various species do not usually mix because they are in equatorial waters at different times of the year.

7

'The Dolphin was in great repute amongst the ancients . . . . How these absurd tales originated it is impossible even to conjecture; for the Dolphins certainly exhibit no marks of particular attachment to mankind. If they attend on vessels navigating the oceans it is in expectation of plunder, and not of rendering assistance in cases of distress.'

Now that we have been able to keep captive dolphins and study their behaviour it seems likely that the ancients, who, after all, lived in countries surrounded by dolphin-filled seas, were nearer the truth than the later armchair scientists.

Even the larger whales have our sympathy. Leviathan no longer seems the embodiment of evil portrayed in the Bible. Few of us would say with Captain Ahab of Herman Melville's *Moby Dick* 'I see in him outrageous strength with an inscrutable malice sinewing it.'

Of course Captain Ahab chased his whale in an open boat, and killed it with hand-thrown harpoons and knives, and the chance of a man being killed was quite as great as the death of the whale. Nowadays we can afford sympathy; we chase our whales from motor-boats with harpoon guns and explosive-headed harpoons which blow holes in the whales as big as rooms, and the whales are then cut up and processed in factory ships in half an hour.

Now our fear is not of the great whales but for them. How many of the species will escape extinction? To our great grandchildren the great whales may have the same sort of reality as has the *Brontosaurus* in the Natural History Museum. I talk of *great* whales to distinguish them from the dolphins and porpoises, killer and pilot whales.

There is a great amount of confusion and many misconceptions about whales. First of all, whales are not fish, although early works refer to them as such. Whales are mammals; they belong to the same Class as ourselves. They breathe air as we do. They give birth to live young (rather than laying eggs), keeping those young inside their bodies during their development and feeding them through an umbilical cord and a special pad of tissue developed for this purpose and called a placenta. When these young are born they are fed on milk produced by mammary glands. The hair which is present in all other mammals would be useless to the whale, which spends its life from birth to death in water, and so it has been dispensed with, and is found only as a few bristles on the snouts of embryo whales and the adults of some species, when, according to Melville, this small moustache gives a 'rather brigandish expression to his otherwise solemn countenance'. Hair is of no use in water, for it is a device used by warm-blooded land animals to retain heat, and keep an unchanged body temperature by trapping a layer of warm still air between the hair and the skin. The whale is, of course, a warm-blooded animal, with a temperature of about 95.9°F, and it keeps this temperature by the abnormally thick layer of fat, known as blubber, under its skin.

Secondly, the animals known as dolphins and porpoises are whales, though small ones. There is little difference between dolphins and porpoises; they are close cousins, but can be told apart by the 'beak' on the front of the dolphin's face. The porpoise has a rounded face and a flat mouth, and is also slightly smaller and slimmer. Some confusion is caused here by the fact that Americans sometimes refer to dolphins as porpoises. Another confusing fact is that there is a fish known as the dolphin fish or dorado *(Coryphaena hippurus),* presumably so called because of its rounded forehead. It is a large, brightly coloured tropical fish and is said to change through all the colours of the rainbow as it dies. But dolphins can be easily distinguished from fish because, as well as having blowholes in their foreheads and lacking gills, dolphins, and all other whales too, have a tail which is arranged the other way round. Fishes' tails lie vertically in the water and move from side to side while whales have horizontal tails that they beat up and down.

Scientists divide whales into two main groups, according to their method of feeding. The largest group, the toothed whales (Odontocetes), which feed on nothing smaller than fish, are mostly small by whale standards. The group contains, amongst others, the dolphins and the porpoises, and the killer whale, whose teeth caused such consternation in Scott and Ponting, and one real giant, the sperm whale.

Paradoxically, the second group, the baleen whales (Mysticetes) are nearly all enormous. The blue whale, at 100 feet long, is the largest animal that has ever lived, but it feeds on minute floating shrimp-like animals known collectively as plankton.

When you visit a dolphin show the animals you are most likely to see are bottle-nosed dolphins *(Tursiops truncatus).* There are another

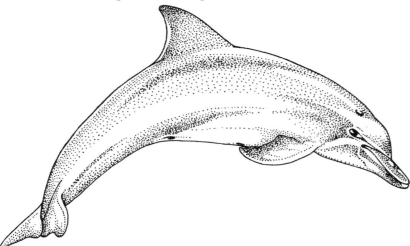

Bottle-nosed dolphin (*Tursiops truncatus*).

forty-six species of dolphin, but the bottle-nosed dolphin is the species most commonly kept because it has a world-wide distribution and is, therefore, comparatively easy to obtain, and it does well in captivity, perhaps because its normal habitat is shallow seas.

Bottle-nosed dolphins are extremely winning animals, very intelligent, curious and friendly, with a fixed 'smile', which is given them by the blunt beak or snout and the curled edges to the mouth. This beak has given the animal a number of names over the years. The Greeks nick-named it Simo which means 'snub-nose', while the medieval French referred to it as a goose, calling it Oie or Bec d'Oie, but ate it on Fridays, conveniently counting it as a fish, and not as a bird, nor as a mammal. Considering that they also looked on the beaver as a fish to be eaten on Fridays, it is perhaps not altogether surprising. The name 'bottle-nosed dolphin' was given to it by American sailors since it reminded them of a certain type of gin-bottle. Flipper, who is the star of a popular American television series, is a bottle-nosed dolphin.

The other well-known species of dolphin is the common dolphin (*Delphinus delphis*). This animal has a sharper beak than the bottle-nosed dolphin and its black and white body is streaked with yellow and brown, whereas most dolphins are solely black and white – as though dressed for a formal occasion. The common dolphin is a deep-water animal and takes less happily to captivity, and so is

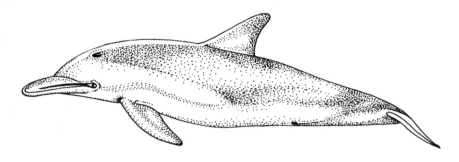

Common dolphin (*Delphinus delphis*).

rarely seen in dolphin shows. This is the animal usually seen playing around the bows of ocean-going ships. Taking advantage of the bow wave it can coast in the forward-moving pressure field in front of the ship. Melville calls it the Huzza porpoise:

> 'I call him thus, because he always swims in hilarious shoals, which upon the broad sea keep tossing themselves to heaven like caps in a Fourth of July crowd . . . . They are the lads that always live before the wind. They are accounted a lucky omen. If you yourself can withstand three cheers at beholding these vivacious fish, then heaven help ye; the spirit of godly gamesomeness is not in ye.'

Both these species of dolphin are sociable animals and live in schools of up to 1,000 individuals, moving about the seas with the shoals of fish on which they feed. These schools consist of males and females with their young, and probably within the school are a number of family groups. Certainly in captivity dolphins do best in family groups consisting of one male and one or two females plus their dependent young. If more than one male is present, or no male at all, then quarrels and fights break out.

The distribution of some dolphin species is much more local than that of the bottle-nosed and common dolphins. Some keep to the tropics and some to the cold sub-Antarctic seas, while others prefer estuarine waters, and there are even four species of fresh-water dolphins.

These are curious, rather primitive dolphins, with a definite neck region and a poorly developed dorsal fin, which, in other dolphins, is large and hooked, rather like the fin of a shark. The susu or Gangetic

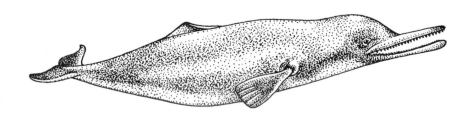

Susu or Gangetic dolphin (*Platanista gangetica*).

dolphin *(Platanista gangetica)* has five-lobed front fins, as though it had not quite lost its fingers. It is also more or less blind, having lensless eyes; but, living in such muddy waters, sight would probably be useless to it.

The Amazon dolphin or bouto *(Inia geoffrensis)* also lives in a world of mud. It has very tiny eyes and finds its way about with the aid of short sensory bristles distributed all over its snout. Its success in

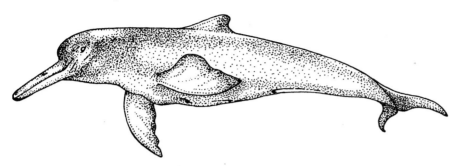

Bouto or Amazon dolphin *(Inia geoffrensis)*.

this may be judged by the fact that it kills and eats the infamous piranha fish, with the result that Amazonian fishermen know themselves to be safe in the water if the bouto is near.

Porpoises, which are mostly coastal species, seem also to like the water of estuaries and rivers, and have been caught in the rivers Seine, Rhine and Meuse, while one species, the Indian or finless black porpoise

Common porpoise *(Phocaena phocoena)*.

Indian or finless black porpoise *(Neomeris phocoenoides)*.

(Neomeris phocoenoides), travels up the Yangtse Kiang to the Tung Ting Lake, 1,000 miles inland.

The porpoises and dolphins are the midgets of the whale world, porpoises being five to six feet long, and dolphins about nine to fourteen feet, according to their species. But there are a number of larger toothed whales, such as the little-known beaked whales, the Arctic-living beluga or white whale and narwhal, and the pilot and killer whales. Melville, in a rather eccentric classification, classes these as Octavo whales, whereas dolphins and porpoises are Duodecimo whales and the great whales are Folio whales.

The beaked whales or bottle-nosed whales (family Ziphiidae) are between fifteen and thirty feet long. Not a great deal is known about them, since they are deep-water animals with little economic importance, and have only been studied when the occasional animal

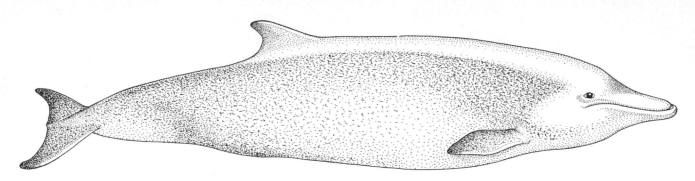

Cuvier's beaked whale (*Ziphius cavirostris*).

has become stranded on shore. They feed mostly on squid, not fish, and are peculiar in having a reduced number of teeth; some have no teeth at all, and some only a pair of small tusks in the lower jaw.

The narwhal *(Monodon monoceros)* also shares this peculiarity, having no teeth in its jaws, but a long, twisted, hollow tusk sticking

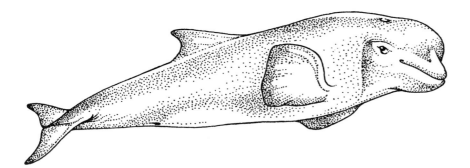

Narwhal (*Monodon monoceros*).

out from the end of the nose for as much as eight or nine feet. Only the males have this tusk, which is developed from one or two of the tooth-buds on the left-hand side of the jaw. The use to the animal of this magnificent appendage is obscure; it is neither used to stir the mud for fish, nor, as has been suggested, to break the Arctic ice, nor apparently for fighting or defending itself. Biologists are left only to suggest that it is a secondary sexual characteristic used to attract female narwhals, in the same way that a beautiful tail will attract a pea-hen, or a fine beard or moustache may attract some women. Narwhals are found only in Arctic waters, where they are hunted by Eskimos, not, as one might expect, for their meat or blubber, but for their skin. This is chewed raw as it contains an extremely high proportion of Vitamin C, a vitamin obtained in less harsh environments from green vegetables and citrus fruits.

Above The head of a beluga. This is the only species of whale that is completely white, except for the rare albino individuals of other species. When very young the belugas are dark grey. This colour phase then gives way to a mottled grey one, followed by a yellow one, before the creamy white coloration of the adult is attained. Belugas are only found in the Arctic.

Left Beluga or white whale (*Delphinapterus leucas*).

The beluga or white whale *(Delphinapterus leucas)* is also an Arctic animal. Like the narwhal it is about eighteen feet long, but its colour is unique amongst whales. It is also the most vocal of the whales; or at least it is the whale most easily heard by man, and this has earned it the charming name of 'sea canary'.

The pilot whales *(Globicephala* species) and the killer whale

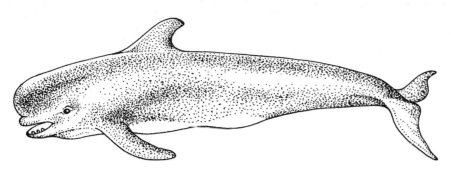

(Orcinus orca) are somewhat larger, being between twenty and thirty feet in length. Both types are found in schools, although in the case of the killer the group is sometimes called a pack, for, as we have seen, killer whales hunt like wolves and will eat mammals as well as fish

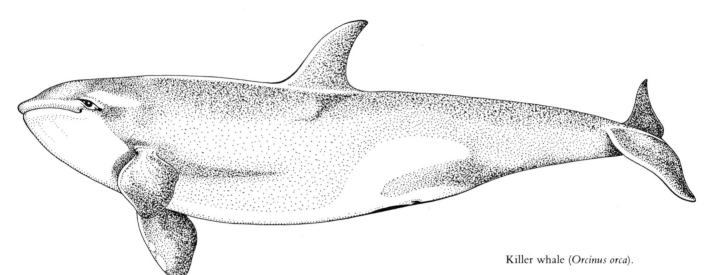

Killer whale (*Orcinus orca*).

and squid, and, though I cannot find any account of a man being eaten, they certainly eat seals and dolphins and will even attack large whales such as the grey and blue whales. The killer whale is found in all oceans from the Arctic to the Antarctic.

The various species of pilot whale form a cosmopolitan genus, but they make seasonal migrations, moving with the squid on which they feed, into colder waters in the summer and warmer waters in the winter. This is a common migration pattern, as we shall see, in the great whales, and it may be that some of the smaller toothed whales already mentioned follow this pattern too, although nobody has yet worked out the migration movements of these species.

Schools of pilot whales can be enormous and have a definite leader (hence the name 'pilot'), and large numbers are sometimes found stranded on shore. Indeed this stranding seems to happen to all toothed whales at various times but why it happens is not really very clear. Nevertheless, hungry men have always taken advantage of this, though to a well-fed nation a school of stranded whales can be quite an embarrassment. The people of the Faroe Isles make the most of the pilot whales' implicit following of the leader and drive them towards the shore. Once the leader has become stranded all the others follow him on shore where they are butchered.

There is only one great whale amongst the toothed whales and that is the sperm whale *(Physeter catodon)*. It is a giant of sixty feet, with

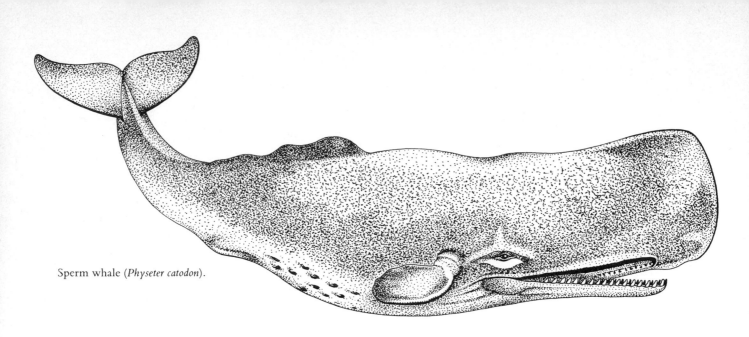

Sperm whale (*Physeter catodon*).

a great square head and a long narrow lower jaw studded with conical teeth which do an efficient job of holding the squid on which it feeds. The teeth of all toothed whales are conical, not being used for chewing but merely for preventing the escape of fish and squid, which are then bolted down whole.

The gigantic square forehead of the sperm whale is filled with an

A false killer whale displays a magnificent mouthful of teeth. Like the killer whale, the false killer has an almost universal distribution.

extremely pure, fine oil which was found invaluable in the nineteenth century for lubricating the newly invented steam-engines and is still used on machinery to this day. This oil gave the whale its name, for when it was first marketed it was so much purer than the oil from the right whale (which was the whale most usually hunted until 1712, when the first sperm whale was killed – by mistake) that it was popularly thought to be whale semen and therefore called spermaceti, which means 'whale's sperm'. Thus the whale that provided it became the spermaceti or sperm whale.

Sperm whales, seemingly unlike other whales, do not usually live in mixed schools but in harems where one large male steer leads a large group of cows and calves. These schools are found in tropical and sub-tropical waters in both the Pacific and Atlantic Oceans. This is probably because warm waters are suitable for the birth and the raising of the calves. There are restricted seasonal movements of these herds, to warmer water in winter and north and south to cooler waters in the summer, as they follow the squid. The young males, once they have left their mothers but before they have fought for and won a harem of their own, live in bachelor schools which, in the summer months, fish the waters of the Arctic and Antarctic which are rich in fish and squid. These young males return to tropical waters in the winter, however.

The literature of whaling is full of grisly stories of the havoc caused by lone rogue sperm whales. One of these tales of a terrifying whale called 'Mocha' Dick was immortalized by Herman Melville, with a rechristening of the whale, in his famous novel *Moby Dick*. Perhaps these lone bull whales are deposed leaders of harems.

The whale-bone whales have no teeth, but feed by filtering out immense numbers of small shrimp-like creatures, known as krill, with their baleen. The fringed inside edges of the baleen plates, which grow down from the roof of the whale's mouth, show well in this photograph. These fringes form a very efficient sieve.

The other giant whales all belong to the second group of whales, the baleen, or whale-bone, whales. They differ from the sperm whale and other toothed whales chiefly in the structure of their mouth. They are toothless but on either side of the vast mouth is a row of horny plates descending from the roof. These are arranged parallel to each other and at right-angles to the length of the whale; on their inner edges they are frayed to form a highly efficient sieve. This enables the whales to filter out from the water the tiny animals and plants known as plankton, krill or brit.

Plankton is a mixed bag of tiny animals, none longer than three inches, and the even smaller one-celled floating plants on which they live. A large number of these planktonic animals are crustaceans – either copepods such as the water-flea or shrimp-like euphausians. Plankton is present in such vast quantity, particularly in cold seas, that it colours the water, which has been described as 'boundless fields of ripe and golden wheat', and also, less lyrically, as 'looking like brown Windsor soup'. All the whale has to do is open its enormous mouth, and in flows the plankton with the sea-water which is then forced out of the sides of the mouth between the baleen plates by the action of the

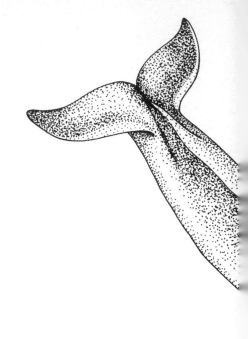

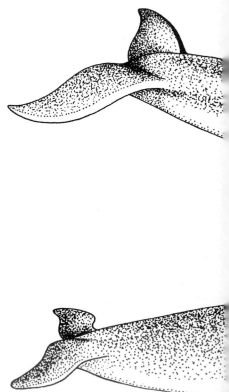

A sperm whale surfaces. The sperm whale has a single blow hole, like all toothed whales, but unlike that of the others it is not central, being displaced by the spermaceti-filled case to the left side of the head.

tongue (which itself can weigh more than a small elephant). The plankton is retained in the mouth by the great doormat of baleen and is then swallowed. Thus the baleen whales browse on these vast water meadows like huge cows. At times the stomach of a large whale has been accidentally split open during the dismemberment of the animal, on the deck of the factory ship, and more than a ton of plankton has gushed out.

The largest of all whales is a baleen whale. This is the blue whale *(Balaenoptera musculus)*, a giant of 100 feet and 160 tons or so. Even the newborn calf weighs more than a full-grown elephant. This whale belongs to a group of whales called rorquals or fin whales, which are characterized by having a fin on the back and long parallel grooves on the underside of the throat. These are thought to open up like accordion pleats, when the animal is feeding, to increase the volume of the mouth. The blue whale was the greatest prize for modern whalers, often producing 120 barrels of oil. It has been brought almost to extinction by them.

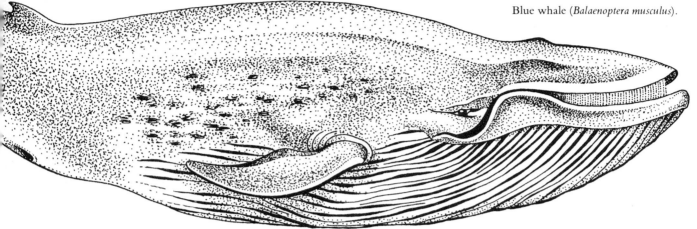

Blue whale *(Balaenoptera musculus)*.

Next to the blue whale in economic importance is the fin whale or common rorqual *(Balaenoptera physalus)* which reaches a length of up to eighty feet. If the whaler can find none of these he will hunt the sixty foot long sei whale *(Balaenoptera borealis)*. As the stocks of each of these species have been depleted the whalers have turned to smaller and smaller whales, and now hunt the smallest of the fin whales, the minke whale or lesser rorqual *(Balaenoptera acutorostrata)*, a mere thirty feet in length.

Fin whale or common rorqual
(Balaenoptera physalus).

Sei whale *(Balaenoptera borealis)*.

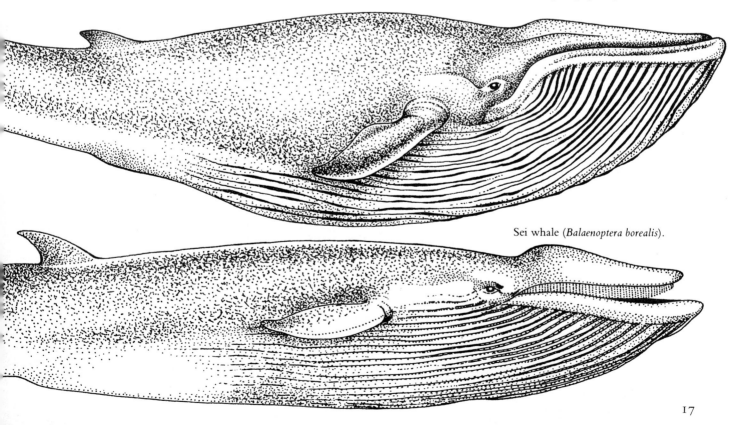

When kept in captivity, the killer
whale is so friendly towards man that
the trainer can safely put his head
into the animal's mouth, with its lethal
teeth, during a dolphinarium
performance.

Left Two dolphins ride in the bow wave of a ship, taking it in turns to get into the forward pressure field, created by the ship, where they can 'free-wheel'.

Below Dolphins and whales being mammals have to come to the surface to breathe. The top of the head breaks the surface first and the blow hole opens, the animal blows and then inhales quickly before the strip of back disappears from view in a smooth wheel-like motion as the animal dives below the surface. The blow of a dolphin is diffuse and not as easily seen as the blow of the great whales, but on still days it can be heard as a sigh.

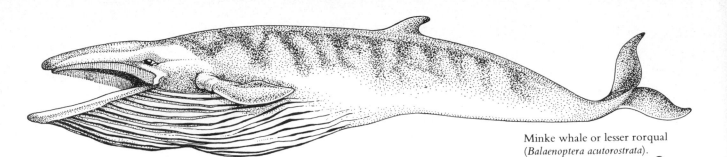

Minke whale or lesser rorqual
(*Balaenoptera acutorostrata*).

All these whales are beautifully streamlined and very fast-moving. They were quite beyond the reach of the old-style hand-harpoon whalers and were shunned by them. For if you harpooned one of these by mistake – and it is not always easy to identify a whale when all you can see above the water is its spout and a small strip of its back – your open boat was likely to be dragged off at such speed on a 'Nantucket sleigh ride' that you finished up out of sight of the mother ship.

The humpback whale *(Megaptera novaeangliae)* which is also a rorqual is not nearly so streamlined or so fast-moving. Its body is bulbous and its pectoral fins are very long, almost a third of its body length, and they have a scalloped appearance.

Humpback whale (*Megaptera novaeangliae*).

All these whales are found in the northern and southern hemispheres, although the northern and southern strains of each species probably do not mix much. They make seasonal migrations, swimming either to Arctic or Antarctic waters in the warmer months but moving back towards the equator as the weather gets colder and their source of food becomes frozen over. For some species the migratory routes have been worked out, but for others they are less clear, and it would seem that some sei whales do not migrate at all but stay off the Norwegian coast all year round. Another common characteristic seems to be that they all occur in family groups which may unite into much larger schools where food is abundant.

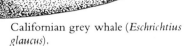

Californian grey whale (*Eschrichtius glaucus*).

Biscayan right whale (*Eubalaena glacialis*).

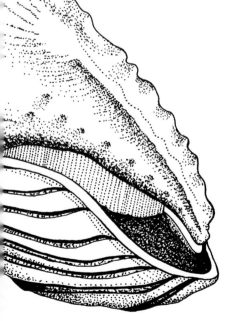

The remaining families of whale-bone whales have no back fin and very few or no throat grooves. The Californian grey whale (*Eschrichtius glaucus*) has the most restricted distribution of all great whales, feeding in the Arctic in the summer months and migrating down to Korean waters and to the coastal lagoons of California. This vulnerability brought the species to the verge of extinction but it is now legally protected and its numbers are increasing. Nowadays, thousands of people gather on the cliffs of San Diego in December to watch these whales pass.

It seems likely that at one time there were also grey whales in the Atlantic Ocean, since parts of their skeletons have been dug up in the sands of Holland. Probably their habit of breeding so close to land led to their extinction in this ocean, although there is no historical evidence of this.

The right whales, or bowheads, of which there are several species, are the whales which look as though their mouths have been put on upside down. Their heads are enormous, being nearly a third of the length of their body, and their baleen is exceptionally long – sometimes the plates are as much as fourteen feet in length. In the eighteenth and nineteenth centuries this whale-bone was very valuable, as it was used for the bones in ladies' dresses, corsets and stays.

This fact, together with their inability to move quickly, made these whales the most hunted ones before the invention of the

Above A pilot whale (or black fish) swims near the surface of the water. Here the light rays penetrate the water but at greater depths the sea is quite black, so that the deep-diving species such as the sperm and bottle-nosed whales feed in total darkness. Darkness, however, is of no consequence to toothed whales which can find their way around with the aid of their 'sonar'.

Above right The beluga, like other toothed whales, has a single blow hole, whereas the whale-bone whales have a double blow hole which looks much more like the nostrils of land mammals.

Right Pilot whales usually swim in large schools, some of which may contain several thousand individuals. The school has a leader, which is usually a male, and this has given the species its common name.

The killer whale, in the wild, is a notorious killer, feeding largely on warm-blooded creatures such as other whales, seals, and penguins which it may first have to knock off the ice. Penguins have been known to take refuge from these animals aboard Antarctic whalers or even among the dead whales that are being towed by the vessel.

harpoon gun in 1840. They were named right whales by whalers because, quite simply, they were the right whales to catch. In fact, apart from the slow humpback and sperm whales, they were the only whales that they were *able* to catch.

Like the other baleen whales they feed in cold waters in the summer and migrate to warmer waters in the winter. There are species of right whale in both hemispheres and in the Atlantic and Pacific Oceans, and the earliest accounts we have of whaling, in the twelfth century, were of the Biscayan right whale *(Eubalaena glacialis)*.

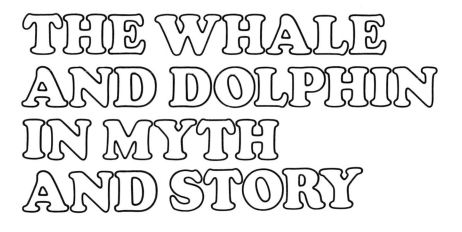

THE WHALE AND DOLPHIN IN MYTH AND STORY

Diviner than the dolphin is nothing yet created; for indeed they were aforetime men and lived in cities along with mortals, but by the devising of Dionysos they exchanged the land for the sea and put on the form of fishes.
Oppian (Halieutica)

Many years ago a small elegant sailing boat was making its way slowly under an intense blue Mediterranean sky between the Greek islands of Ikaria and Naxos. The prow of the boat was carved like a fish and the arching stern like the head of a goose, or perhaps both were like dolphins. The sail hung and fluttered in a faint wind and the sailors bent their brown backs over the oars. But all was not as idyllic as it seemed, for these sailors were planning to abduct their passenger and sell him into slavery. He was obviously well-born and rich, but what the villainous crew did not realize was that he was Dionysos, the Greek God of Wine and Frenzy (later named Bacchus by the Romans).

When Dionysos realized their treachery he began to confound the sailors with magic; he turned their oars into snakes and filled the ship with vines and the sounds of flutes. The terrified sailors dived into the sea to escape this madness and were transformed into dolphins by Poseidon (Neptune) the God of the Sea.

Thus, according to Greek legend, dolphins were originally men, and this explains the rapport felt between man and these animals. This legend can be seen depicted on the Dionysos cup which is still intact although it was made 540 years before the birth of Christ. Indeed, dolphins abound in Greek legends and art, being found in murals and mosaic floors, on coins and statues.

Dolphins are often depicted accompanying Aphrodite, Goddess of Love, who rose from the foam caused when pieces of her mutilated father were thrown into the sea by her murderous half-brother Cronos. Cronos had no reason to love his father who had banished his children from birth to a land as far below Hades as Heaven was above Earth. But then, neither was Cronos an ideal father, for he ate his own children.

Poseidon too was usually shown with dolphins, which often drew his sea-chariot, and it was he, according to legend, who put the

Left The Dionysos Cup (540 B.C.) probably illustrates the Greek myth which accounts for the first appearance of dolphins, when villainous sailors attacked the god Dionysos and were then turned into dolphins.

Above Dolphins are depicted in a mural from Knossos. These compare well with mediaeval drawings of dolphins which usually showed them with scales, gills, fins and even more bizarre organs.

dolphin constellation in the sky where it can be seen in July. He did this out of gratitude to the dolphins for finding him his bride, Amphitrite, who was hiding from him in a sea-cave. Later he had further reasons for gratitude to the dolphins since they rescued his son, Taras, from drowning.

To the Greeks, and to the people of the Mediterranean lands where Greek culture spread, the dolphin became a symbol of swiftness and diligence and love. It became a talisman for voyagers on sea and land, and also for those voyaging into the after-life, so that the dead were buried with dolphin tokens in their hands.

In addition to the legends about dolphins there are a number of stories in Greek writings which are probably at least partially true. These stories, told by Herodotus, Plutarch, Pliny the Elder, Oppian and Aristotle, are the ones that came into such disrepute in the last

A mosaic of a dolphin from 'The House of Tridents' in Delos. Dolphins abound in ancient Greek art.

28

century. But their stories of dolphins befriending children whom they allowed to ride on their backs, and of life-saving rescues, and human bodies brought to shore by dolphins have been paralleled so accurately, during this century, that we can no longer write off the Greek stories as merely sentimental fables.

Let us look at the stories of rescue first. Taras, the son of the sea-god, has already been mentioned, and Telemachos, son of the most famous adventurer of all time, Odysseus, is said to have been rescued in the same way, and for that reason Odysseus had a dolphin emblazoned on his shield and ring.

Arion, a famous poet, musician and singer of his day, who was born on the island of Lesbos 600 years B.C., no doubt knew of these rescues and the legend of Dionysos and the dolphins. Perhaps he merely amalgamated them to make a poem to sing as he accompanied himself

Silver coins from Syracuse (466 to 430 B.C.) show the head of the nymph Arethusa surrounded by dolphins. Dolphins were a favourite motif for ancient coins. The dolphin was supposed to provide safety for travellers and so a representation of a dolphin had the same significance as a St Christopher medallion has today.

A bouto (left), kept in captivity with bottle-nosed dolphins. These boutos, which live in the Amazon and its tributaries, are considered sacred by the Amerindians and are surrounded by myths just as were the dolphins that swam around the shores of ancient Greece and Rome.

A white-sided dolphin. Throughout
the world dolphins are considered to
be the friends and protectors of
sailors.

on the harp. Certainly his dolphin story bears a striking resemblance to the Dionysos legend, but, who knows, it may be true. Here is the story.

Arion, after a successful tour of Italy and Sicily, and loaded with money and prizes, took ship for Corinth. He chose a Corinthian ship rather than an Italian one for he trusted the Corinthians more. But evidently sailors were an untrustworthy lot, for very soon they were plotting to kill him and keep his treasures. Arion begged for his life, but they told him that he must either jump overboard or die by his own sword if he wished for a proper burial ashore. As a last favour Arion pleaded to be allowed to sing, and, dressing himself in all his splendid clothes and weighed down in his riches, he stood in the stern and sang them the 'Orthian', a high-pitched song addressed to the gods, and as he finished he leapt fully clothed into the sea.

A dolphin, perhaps attracted by the shrill sounds, took Arion on its

The Poseidon relief from Thera on the Greek island of Santorini. Poseidon, the Greek god of the sea (re-named Neptune by the Romans) is usually portrayed with dolphins. He had good reason to be grateful to them for they not only found him his bride but also saved the life of his son.

back and swam with him to Tainaron at the southernmost tip of the Greek mainland. From there Arion made his way overland to Corinth to confront and bring to justice the greedy sailors. As a thank-offering he placed a small bronze statue of a man on a dolphin in the temple at Tainaron where it was seen 700 years later by Pausanias, the Greek historian.

From the coast of Turkey, at that time part of Greece, come two other stories of rescue. One is of a grown man, Koiranus, who was saved from death by shipwreck by a dolphin which he had earlier befriended and saved from fishermen who were intent on killing it. The other is of a boy, Hermias, who was washed off the back of a dolphin he was riding and drowned. The body of the boy was brought back to the shore by the dolphin which grounded itself and also died.

In modern times there have been a number of such rescues reported in the newspapers. During the last war a rubber dinghy containing six American airmen was pushed ashore in the Pacific by dolphins. In 1960 a Mrs Yvonne Bliss, who managed to fall unnoticed from a ship in the Bahama Channel, was guided ashore by a dolphin, and in 1966 a bather was saved from sharks and helped ashore by dolphins in the Gulf of Suez. In 1949 a lady quoted by the American Magazine of Natural History as being the wife of a lawyer, and therefore thought to be reliable, was caught by an undertow while bathing off the Florida coast and swept out to sea. She panicked, swallowed water and thought she was drowning when something pushed her violently from behind so that she landed on the beach with her nose in the sand, too exhausted to look around. When she was able to do so there was no one near her, but in the water twenty feet from the shore a dolphin was jumping and swimming in circles.

Perhaps at first glance these reports, old and new, look like wishful thinking. It is pleasant and comfortable to believe in the goodwill of dolphins towards ourselves, but in fact there is some scientific evidence to support the stories. New-born dolphins are pushed to the surface by their mothers. This is an essential maternal instinct in dolphins, since otherwise the young might well drown. Sick dolphins and dolphins in a state of shock after being captured are also aided in this way, as the keepers and trainers at any dolphinarium will testify. Even dolphins of different species will do this for each other in captivity.

It seems probable that the rescuing of human beings is allied to this. Whether the dolphins knowingly rescue men we cannot be sure. They have been seen to lift and escort a mattress to the shore in a similar way, and at Marineland, California, in 1961 a female dolphin lifted a dying five-foot leopard shark to the surface and supported it for eight days. So perhaps this rescue is an instinctive reaction to anything roughly of dolphin size and shape that is sinking in the water. But if this is so why are humans brought back to the shore? This is not what would be done to a sinking dolphin.

Modern instances of friendship between wild dolphins and man are many and well documented. In 1945, Sally Stone, a thirteen-year-old American, palled up with a dolphin at Long Island Sound; in 1953 at Fish Hoek near Capetown two friendly dolphins played with bathers; and from 1960 until 1966 at Elie in Fifeshire and Seahouses in Northumberland a dolphin fraternized with boats and bathers. The two best-known stories, however, come from New Zealand.

Pelorus Jack (all cetaceans seem to be called Jack in the Antipodes)

This Roman mosaic from the second or third century A.D. is rather reminiscent of the Dionysos Cup.

faithfully escorted boats across Cook Strait between Wellington and Nelson from 1888 until 1912 and became so well known and liked that Governor Plunket made an Order in Council that 'It shall not be lawful for any person to take the fish or mammal of the species commonly known as Risso's dolphin *(Grampus griseus)* in the water of Cook Strait or of the bays, sounds and estuaries adjacent thereto.' Governor Plunket does not seem to have been very sure of his zoological facts but his heart was in the right place.

In 1956 a similar law was passed to protect a dolphin at Opononi Beach, Hokianga Bay. This time it was a bottle-nosed dolphin called Opo. It had first been christened Opononi Jack, but this was soon shortened when the dolphin turned out to be a female. She became so friendly that she would allow herself to be partially lifted from the water and caressed; she would throw and catch a beach ball; and she would allow small children to be put on her back, and there are photographs to prove this. Her particular friend was thirteen-year-old Jill Baker, who had the following to say about their relationship: 'The dolphin became so friendly with me because I was gentle with her and never rushed at her as so many bathers did . . . on several occasions when I was standing in the water with my legs apart she would go between them and pick me up and carry me a short distance before dropping me again.'

Opo's fame soon spread; during the summer the nearby camp was full and the hotel booked for months ahead. At week-ends special deliveries of beer and ice-cream had to be made for the thousands who arrived at the beach to see Opo, and the roads around were blocked with traffic. Everyone wanted to touch Opo, and some people got so excited that they would run fully clothed into the water to do so.

Anthony Alpers, the New Zealander who investigated the story of Opo and reported it in his books, *Dolphins* and *A Book of Dolphins*, described that summer at Opononi Beach as follows.

'On this mass of sunburned, jostling humanity the gentle dolphin had the effect of a benediction. Saturdays, for all the crowds and all the beer, were not as New Zealand Saturdays so often are . . . there was no case of drunkenness, fights or arguments. Everybody was in the gayest holiday mood.'

The local tradesmen flourished and there was talk of building a new wing on to the hotel, but before this could be done, the dolphin was found dead. Letters of condolence to the people of Opononi poured in. Sir Willoughby Norrie sent a message of sympathy from Government House, and a statue was constructed to the dolphin's memory.

Compare this factual account with a story told by Pliny the Younger in a letter to his friend, the poet Caninius, in A.D. 109:

'I have a story for you, which is true, though it has all the qualities of a fable, and is worthy of your lively, elevated, and wholly poetical genius. I heard it the other day at table, when the conversation turned on various miraculous events. The man who told it is completely reliable – though what is that to a poet? However, you could depend on his word even if you were writing history.

There is in Africa a Roman settlement, near the sea, called Hippo. It is beside a navigable lagoon, and from this lagoon like

a sort of river there runs an estuary, whose waters are carried out
to sea and back again with the tides. The inhabitants of all ages are
very fond of fishing and boating, and of swimming too; especially
the boys, who love to idle their time away in play. Their idea of
glory is to be carried out to sea, the winner being whoever leaves
the shore and his rivals furthest behind. In a contest of this sort one
lad who was bolder than the rest was getting far out, when
suddenly a dolphin came up to him. First it swam in front of him,
then it followed him, and then went round him. Then it dived
under him and took him on its back, and rolled him off again, and
to his horror started carrying him out to sea. But then it turned back
and restored him to his companions and dry land. The news of this
got around, and everyone rushed to see the boy, as if there were
something supernatural about him. They eyed him and questioned
him, listened to him, and passed his story around.

Next day they lined the shore, gazing out to sea and watching
the lagoon. The boys went swimming, and among them our hero,
but with more care this time. The dolphin duly appeared, and made
for the boy, who fled with the rest. At that the dolphin, as if
inviting him back, started leaping out of the water and diving and
twisting and twirling about. This happened again the next day,
and on a third day, and for several days, until the men of Hippo,
born and brought up by the sea, began to be ashamed of their fears.
They went up to the dolphin, and played with it and called to it.
They even touched it, and it encouraged them to stroke it. With
experience, they grew venturesome. In particular, the boy who had
had the first encounter with it swam beside it in the water, and
got on to its back, and was carried to and fro. Feeling that the
dolphin knew him and was fond of him, he became fond of the
dolphin. There was now no fear on either side, and the boy's
confidence and the dolphin's tameness increased together. Other
boys, too, swam to the right and left of their friend, urging him on
and telling him what to do. What is also remarkable, another
dolphin accompanied this one, but only as an onlooker and escort.
It did none of the same things, and submitted to none of the same
familiarities; it merely conducted the other one to and fro, as the
other boys did with their companion.

Believe it or not (but it is just as true as what I have described)
this dolphin that played with boys and gave them rides would also
come out on the beach and get dry on the sand, and when it had got
warm would return to the sea. It is known that on one of these
occasions the Governor's legate, Octavius Avitus, moved by some
stupid superstition, poured some precious ointment on the dolphin.
To get rid of the strange feeling and the smell, it made for the open
sea, and was not seen again for several days, and then it seemed
languid and sickly. However, it soon recovered its strength, and
became as playful as before, and was giving rides again. The sight
attracted all the officials of the province. But the expense of
welcoming and entertaining them during the stay began to exhaust
the modest resources of the town; besides, the place itself was
beginning to lose its quiet and peaceful character. So it was decided
to do away secretly with the cause of the invasion.

How tenderly, how imaginatively, you will tell this sad tale.
What colour you will give it! How you will heighten its effect!

Ancient Greek and Roman writings
abound in stories of friendships
between boys and dolphins. In the
twentieth century dolphins have been
known to single out children as
companions.

Yet there is no need for you to add anything to it. It will suffice merely to make sure that the truth speaks for itself. Farewell.'

The only real difference in these two accounts is in the rules of hospitality in twentieth century New Zealand and in the second-century Roman Empire. In ancient Rome hospitality had to be offered without charge and therefore the dolphin became an embarrassment. In Opononi the townsfolk, and therefore the dolphin also, prospered.

There are several similar tales of dolphins in ancient writings. There was the friendship between a dolphin and a boy on the island of Poroselene off the coast of Asia Minor, and another friendship was reported from the nearby town of Iasos, where a boy, Dionysios, was befriended by and rode a dolphin. Alexander the Great got to hear of this and made the boy High Priest of Poseidon at the temple of Babylon for he thought the boy must be especially favoured by this god.

My favourite story comes from the ancient Romans, not the Greeks, and took place at about the time of Christ. Near to Naples is a lake separated from the Mediterranean Sea only by a narrow strip of land and known at that time as the Lucrine Lake. A dolphin had been brought into this lake and given the usual dolphin name of Simo. This dolphin became very friendly with a poor boy who lived on one side of the lake and had to travel all the way round the lake every day to his school on the opposite side. The dolphin would come when his

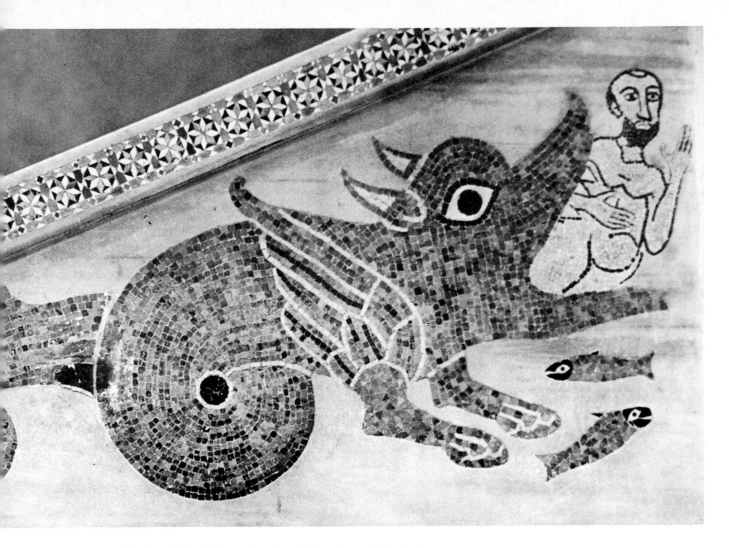

name was called and feed from the boy's hand, and finally became so friendly that the boy took the short cut across the lake to school every day, riding on his back. The story has a sad ending unfortunately, for the boy fell ill and died, and the dolphin did not outlast him long, dying, according to Pliny, 'purely of sorrow and regret'.

It seems strange that there are virtually no accounts of friendly dolphins in Western literature between the second and twentieth centuries, but perhaps the reason lies in man's behaviour rather than in the dolphins'. Dolphins are intelligent creatures and it takes them very little time to realize what their reception is going to be like. The ancients considered it a sin to kill a dolphin (although if driven by hunger or cupidity they would still do it). This is reflected in the words of the poet Oppian: 'equally with human slaughter the gods abhor the deathly doom of the monarchs of the deep'. Today also, most people will not kill dolphins, although 200 thousand are killed accidentally every year in the tuna fishing industry, and perhaps whalers will turn to killing them when they have exterminated the larger whales.

However if we turn to the literature of other parts of the world we find that the idea of a dolphin as a friend and as a rescuer of sailors is universal.

In South America the bouto of the Amazon are considered sacred and must not be killed. They are said by the Indians to be very fond of young girls and to disguise themselves as humans at carnival time.

If an unmarried girl becomes pregnant it is believed that the bouto is responsible and this is not considered any disgrace.

The Vietnamese consider dolphins and whales to be sent by 'the god of the waters' to protect sailors and bear up shipwrecked men and boats in difficulties. If a dead whale or dolphin is found stranded it is buried and the finder takes the place of its eldest son and mourns it. For three months and six days he must wear a mourning turban, then the turban is burnt and the bones dug up and put into a sanctuary, where an offering is made annually. The Vietnamese believe that there is rain for three days when a whale dies, and therefore in bad weather dead whales are diligently sought and buried if found. Once the whale is buried the bad weather will stop.

The Japanese have a strange ambivalent attitude to whales for, while nowadays their whaling fleet is responsible for a large proportion of whales killed, in the island of Saikai-to, off West Japan, a requiem service is held each April for the whales killed many years ago by the net-whalers. A tomb was built on this island where all the embryo whales found in the slaughtered females were buried, and here a record is kept of them, each one being posthumously given a Buddhist name.

Whales also feature in other ancient writings – in the Bible, both as Leviathan, the monster of the deep, who seems to be the embodiment of evil, and also as the more friendly 'great fish' that swallowed Jonah and regurgitated him on to the land at God's command. They also appear in the fables of the Maoris and in the medieval writings of Scandinavia and Iceland. The Maoris had no written language until the Europeans came to New Zealand, but the legends passed down by word of mouth include the story of a whale named Big Gamboller, which a chief gave to his son as a mount. There are also many stories of legendary gentle sea-beasts called 'taniwha', which Anthony Alpers identifies as dolphins. Certainly the tales of taniwha strike a familiar note. There are stories of men rescued by taniwha and there is a tale which is very like the Dionysos legend, where Ruru the murderer is turned into a taniwha and destined to live for ever by the coast and meet every canoe that passes.

Just as the people of New Zealand and Australia were familiar with the whales of the southern hemisphere, so the peoples of Iceland and Scandinavia knew the whales that visited the Arctic Circle. In fact the thirteenth century Norwegian work *Speculum Regale,* which is full of information about whales, tells us that few things in Iceland are worth talking about except its whales. So numerous were these whales that numbers of Scandinavians settled in the country, attracted there by the whaling. However, there seems to have been little mutual respect for, according to the Icelandic saga *Heimskingla*, the Danish King, angry because the Icelanders had been composing libellous verses about him, sent a magician disguised as a whale to spy out the best place for an invasion.

The Icelandic sagas also contain more factual accounts of whales and whaling. The whaling was primitive and dangerous, and many harpooned whales escaped, later to die and drift ashore. So we find laws passed as early as 1281, relating to the ownership of such whales. The accounts of whales in these works were mostly accurate and compare well with other medieval writings where the descriptions of whales and the illustrations that went with them became more and

The terrors of the sea as they appeared to Sebastian Münsters in 1544. This print from a wood carving appeared in *Cosmographia*, one of the earliest geographical works. Amongst the giant lobsters, fish, and sea snakes are some creatures which are no doubt supposed to be whales. These are depicted with scales, gills, tusks, and funnel-like blow holes. The pig with tusks and a fish's tail is probably meant to be a dolphin. Mediaeval artists drew from verbal descriptions, and highly coloured ones at that, not from actual animals.

more fantastic. There follows an extract from the *Speculum Regale*, describing the Atlantic right whale most accurately, although the conclusions drawn from the facts are a little odd.

'. . . but this fish lives cleanly, because people say it does not eat any food except darkness and rain that falls on the sea . . . . It cannot open its mouth easily because the baleen that grows there rises up in the mouth when it is opened . . . . It does no harm to ships: it has no teeth and is a fat fish and well edible.'

This work also accurately describes the sperm whale, the killer whale with 'teeth like dogs', and the narwhal.

However, these were superstitious times, and, in addition, a number of strange, evil whales were described – the red whale, the horse whale and the pig whale which sank ships and loved human flesh. It was considered unlucky even to mention their names while at sea, and if a careless sailor inadvertently did this he would have his rations stopped as punishment.

As well as these evil whales there were good whales – the rorquals or fin whales – which would protect ships from the evil whales and also drive fish towards fishermen. This latter belief persisted into the nineteenth century, for when the harpoon gun was invented, and the hunting of fin whales became feasible, the fishermen objected, for they thought it would endanger their livelihood. Perhaps there is a basis of truth in this, for dolphins and killer whales have both been known to help men in this way.

If we look at all the stories, true and fabled, we find that man's affection for the large whales is tempered with terror. But between dolphins and men there seem to be only good feelings. Plutarch sums this up in his *On the Cleverness of Animals*, written some 1800 years ago.

'To the dolphin alone, beyond all others, nature has given what the best philosophers seek; friendship for no advantage. Though it has no need of man, yet it is the friend to all men and has often given them great aid.'

BIOLOGY AND EVOLUTION

The aorta of a whale is larger in the bore than the main pipe of the water-works at London Bridge, and the water roaring in its passage through that pipe is inferior in impetus and velocity to the blood gushing from the whale's heart.
Paley (Theology)

Because of their fish-like shape and their mysterious lives in an element different from our own, people tend to think of whales and dolphins as animals very foreign to themselves. But this is not so. There are far more similarities between cetaceans and other mammals than there are dissimilarities. If you could telescope your neck, stretch your coccyx and lower spine, shorten your arms, grow skin between your fingers, lose your back legs and most of your pelvis, lose your hair and your external ears – in fact streamline yourself – and if your upper lip grew so large as to push your nostrils up on to the top of your head, you would make a very passable dolphin.

In fact, if a whale or a dolphin is dissected it is surprisingly similar to the land mammals. Why, for instance, should a whale have a pelvis? The pelvis is for the attachment of the leg bones, and so it is of no use to the whale; but there it is, small but still recognizable and sometimes with a vestigial leg bone or two attached to it. Very occasionally a whale is found with a pair of bump-like back legs protruding from its body, and those on a whale killed near Vancouver were a yard long! All this is enough to make scientists realize that the remote ancestors of whales were land animals with four legs and a proper pelvis, necessary for movement on dry land.

If we compare embryo whales with those of land mammals we find great similarities – very young embryos are pretty well identical. The whale embryo has a small tail with no flukes and four limb-buds, just like a few-weeks-old human embryo. Even the adult whale retains the bones of its five fingers within its flippers, while the tail flukes and the dorsal fin contain no extra bones but are made up of tough connective tissues.

The embryo whale at various times in its development within the uterus sports rudimentary fingers, hairs on its head, and tiny ear flaps. Also present are external nipples and an external penis, which will later sink below the surface of the skin in order to complete the animal's streamlining. The development of an embryo seems to mirror the stages of an animal's evolutionary history but, although the whale's history becomes quite clear by studying these embryos, there are, unfortunately, very few whale fossils to confirm this. Since most whales are deep-sea creatures it is not surprising that their remains are lost.

At the beginning of the nineteenth century fossil whales were discovered in North America, and then later in Europe, Africa, New Zealand and the Antarctic. These Archaeocetes lived about forty-five million years ago. Some were torpedo-shaped like modern dolphins, and some were longer and more snake-like, but they were all much smaller than the great whales of today, and more primitive. That is to say, they had characteristics which were half-way between those of present-day cetaceans and those of land mammals. Their forelimbs, though they were probably flippers, had a skeleton which probably still bent at the elbow and was far less telescoped than that of a modern whale, whose forelimbs only bend at the shoulder. The pelvis was larger and more recognizable as such. The skull, though long, was more like that of a land animal than that of a whale and the nostrils were only half-way up the snout.

For all that, scientists do not think that these were the direct ancestors of our present-day whales, but merely a side-line – a number of species that were not successful and which became extinct some twenty-five million years ago. Modern whales and these Archaeocetes, they think, evolved from small, primitive land mammals that lived 120 million years ago. These small animals had some characteristics of

Conjectural family tree of the cetaceans to show their relationship to one another and to other mammals. It seems likely that the whale, along with a number of other mammals, evolved from small shrew-like mammals living about 120 million years ago. The mammals which seem to be the nearest relatives to the cetaceans are the artiodactyls or cloven-hoofed mammals.

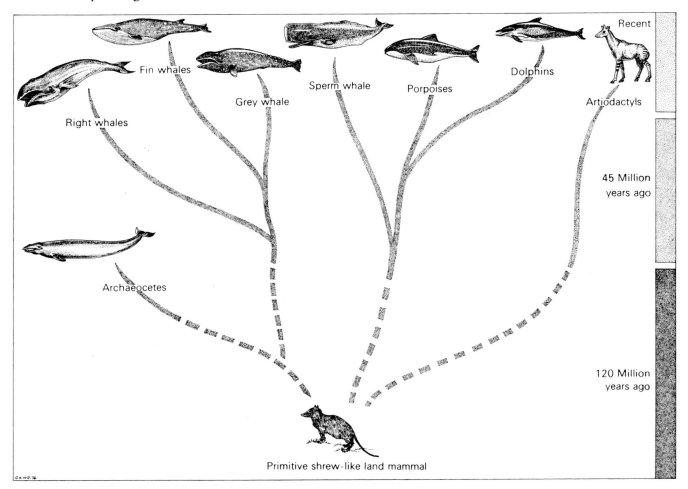

Two common dolphins leap from the sea in formation in Cook Straits, New Zealand. The ability to swim in formation and leap in unison is probably due to the dolphins' 'sonar' as well as their sight.

modern carnivores and insectivores and it is thought that from them evolved not only the cetaceans, but also the carnivores and insectivores, the hoofed mammals, the seals and the sea-cows. If the blood chemistry of all these animals is studied it is found that the blood proteins of whales are most like those of the even-toed ungulates, the group to which the cow, the pig and the deer belong. Therefore, these, unexpectedly, are the whales' and dolphins' nearest relatives. The French name of *porc-poisson,* therefore, which remains with us as *porpoise* is not so inaccurate as one might think.

Nor are the great whales and the dolphins as closely related to each other as one might think. Their body chemistry is different; the whale-bone whale's blubber contains oil, while that of the toothed whales contains a liquid wax. So it seems probable that the two groups diverged very early, perhaps, in fact, developing separately from the early shrew-like land-living ancestors, and grew to look like each other as their bodies were slowly modified to their life in water.

Whales and dolphins are most beautifully adapted to their aquatic life. For instance, they can move through the water at an incredible speed – faster than any fish, and at a greater rate than scientists calculate that they should be able to move. How they do this is still something of a puzzle.

Whales' speeds vary. The slowest are the right whales, with a cruising speed of two knots, although they can, if pressed, make five knots. This slow speed is the main reason that they were the right whales to catch. The grey whale and the humpback are also slow. Sperm whales are faster and have been recorded as travelling at twenty knots, although normally they travel at about half that speed. Blue and fin whales travel at from ten to twelve knots, but they can do thirty knots for a short time. The champion sprinter is the sei whale which can reach a speed of thirty-five knots. Dolphins are surprisingly fast moving for their size. They have been found under experimental conditions to swim at eighteen knots and observers in boats have recorded them moving at twenty knots for some considerable time, and even at thirty knots when swimming in the bow wave of a destroyer. Speeds of fifty or even sixty knots have also been claimed. Thus the dolphins and the faster whales can keep up with ocean liners. However, they are faster by far than submarines which can only achieve a top speed of six knots when submerged. This is a better comparison, for it is more difficult to move under the surface than on it, since the resistance of water is much greater than that of air, as anybody who has tried to run through water knows. It is hardly surprising that scientists are obsessed with the speed of cetaceans – they want to learn their secret.

As the muscles of all animals function in the same way, by comparing the bulk of muscles in any animal we can work out the power that it is capable of realizing by comparing it with the power and muscle bulk of man, which we know. In this way the speed of most animals can be foretold correctly but, given the muscle bulk of a dolphin, it has been calculated that it should only be able to swim at eleven knots, and, as it has been observed swimming at twice this speed at least, either its muscles must be much more efficient than man's, or it must have some way of reducing the resistance of the water. The speed at which a body moves through water depends not only on the energy expended but on the resistance of the water as well, and one of the factors which

affects this resistance is the way in which the water flows over this body. When a boat, for instance, moves through water, the water closest to it tends to cling to it, causing the water to become turbulent, forming little eddies. This causes a drag on the boat and slows it down. Streamlining reduces this drag, but by no means eliminates it. It is thought that perhaps whales and dolphins, in some way we do not fully understand, have a surface that does not cause eddies to form on it, or, in other words, instead of having a turbulent flow of water over them they have what is called a laminar flow. This has never been proved, but if dolphins do have laminar flow it is calculated that, with muscles of the same power as those of human beings, they should move at a rate of seventeen or eighteen knots – their actual speed. An observation made by G. A. Steven of the Marine Biology Station in Plymouth bears out this theory. One evening during the Second World War he saw a number of dolphins and seals swimming in a sea made phosphorescent by millions of tiny one-celled organisms. The dolphins left a wake of two straight glowing lines, whereas the wake of the seals was extremely turbulent.

Many theories have been advanced as to how whales and dolphins manage to achieve this laminar flow. It has been suggested that ridges in the lower layers of the skin may be responsible, or simply that the

The underside of a whale-bone whale's throat, encrusted with acorn and stalked barnacles. The soft skin of all whales is liable to be attacked by various parasites including barnacles, whale lice and lampreys. The right whales, humpback and grey whales are particularly prone to attack by these parasites because of their slow rate of swimming. Blue whales and fin whales tend to pick up a film of tiny, yellow, one-celled plants called diatoms while feeding in the Antarctic. Hence the name 'sulphur bottom' for the blue whale.

up and down movements of the body and tail will cause laminar flow, or that tiny movements in the skin, either inadvertent or controlled by the animal's brain, help to damp out any eddies beginning to form. In fact, Kramer, in 1966, built a model of dolphin skin consisting of two layers of rubber with a network of fluid-filled canals sandwiched in between. By covering a model of a dolphin with this, he showed that it decreased the drag by forty per cent. Perhaps there is truth in all these theories. Perhaps, also, the dolphins' 'secret' can eventually be put to use by the shipwrights. This is what Auguste Piccard foresees in his book *Au Fond des Mers en Bathyscaphe*. According to him we will cross the Atlantic in thirty-six hours, travelling 100 feet below the surface in a liner powered by a small diesel engine and using only as much fuel as two or three lorry engines would. In the rubber covering of the hull will be numbers of pressure gauges feeding information to the ship's computer. This will send out appropriate messages to small magnets which will move the 'skin' of the ship to prevent the formation of eddies. The name of his ship is *The Dolphin*. What else could he call it?

The skin of whales and dolphins is very different from our own, although built to the same plan. Firstly, it has no hairs and no sweat glands, since these are mechanisms for temperature control in air and would be useless to a whale. Then it is very thin, less than ten millimetres in thickness and, according to Melville, 'almost as flexible and soft as satin'. For this reason it is useless as leather (except for

Food web in the sea.

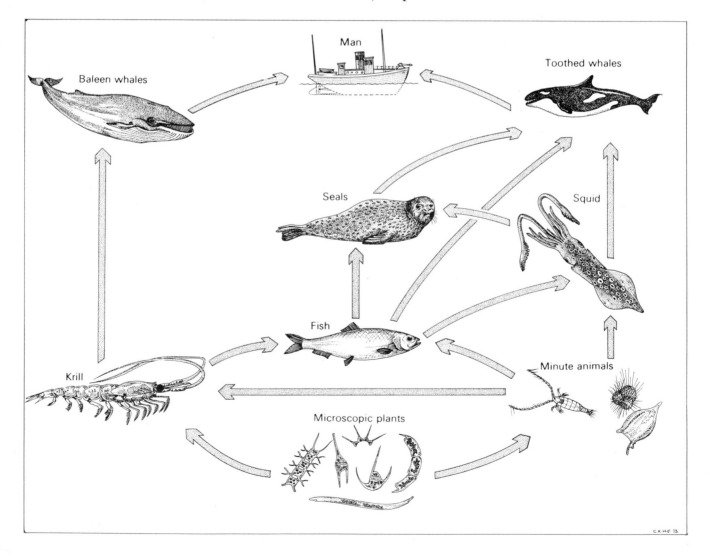

belugas, narwhals and river dolphins, which have tougher hides). The Japanese tried it as shoe leather in the Second World War but found it unserviceable. In the big whales the only skin with any toughness is the skin of the penis, which in the days of the old whalers was stripped off, dried and made into an overall by cutting off the tip and tailoring two arm holes in the right place. This was then worn by the 'mincer', the man who chopped up the blubber into pieces small enough to go into the try-pots to be rendered down. One can imagine that he needed some form of protection from this very greasy job – though who could imagine anything quite as bizarre as this?

Below the skin lies the blubber, which is attached to the muscles beneath by loose connective tissue, so that the blubber can be peeled off the whale quite easily, rather like peeling an apple. This blubber can be up to twenty-eight inches thick and provides a large proportion of the oil for which the whale is hunted. It is therefore the cause of most of its troubles, but it is also a very necessary part to the whale, for without it the animal would die of cold. Water conducts heat twenty-seven times better than air and for this reason a man will die in fifteen minutes in freezing cold water, and will last only three hours in water at $15°C$. Therefore the whale's heat conservation must be good, since many of them spend half the year in Arctic and Antarctic waters. The blubber, in fact, acts not only as an insulator and helps to provide the whale with his smooth, streamlined shape, but also serves as a reserve food store during the winter months when it migrates to warmer waters where the food is not so plentiful.

The body temperature of cetaceans seems to be low, but it is not accurately known for some species, since taking the temperature of a living whale presents problems. In Arctic waters the animals seem to have to move about almost continually in order to keep warm, notwithstanding the fact that large objects lose heat more slowly than small objects (this is the reason why you can eat a small pie taken straight from the oven sooner than a large one). In warmer climates whales can afford to bask and overheating can even be a problem, for a whale can neither sweat nor pant, like land animals, in order to keep cool. For this reason if they are taken out of water they will die of heat stroke, and great care has to be taken, in transporting dolphins, always to keep them wet. If the carcass of a whale is left unopened in warm climates for any length of time the blubber acts rather like the hay in a hay-box and the whale cooks in its own heat. The living whale cools itself by pumping more blood through the blubber to the skin, especially in the fins and flukes, where heat will be lost to the surrounding water.

The raw material that the whale uses to synthesize the fat in its blubber is the plankton or the fish and squid which it eats. It has often been suggested that man should by-pass the whale and extract oil from the plankton, but this oil is not fit for human consumption, and would have to be treated, which would make it a more costly operation. Besides, man is nowhere near as efficient as the whale at finding the concentrations of plankton. Perhaps extracting oil directly from plankton will be feasible in the future.

The whale-bone whales filter the plankton from the sea with their baleen as described in Chapter One. The baleen is a horny substance made out of the same material as our hair and finger nails, and it grows constantly from the roof of the mouth as it is worn away by friction.

The toothed whales' food, however, is primarily fish and squid and therefore they do not have this baleen filter. The whale-bone whales have complicated, many-chambered stomachs, reminiscent of the stomachs of the cow and other ruminants. The intestine too is surprisingly long for a carnivorous mammal; in a large sperm whale, for instance, it is 1,200 feet long or nearly a quarter of a mile!

The throats of the filter-feeding whales are quite small in diameter, about four or five inches, so that if Jonah was swallowed by a whale it must have been either a sperm whale or a killer whale. Killers are supposed to be able to swallow seals and dolphins whole, and a giant squid some thirty-five feet long was once found in a sperm whale caught in the Azores.

One of the problems that the whale comes up against is the amount

The intestines of a sperm whale are surprisingly long for a carnivorous animal. The diet of the sperm whale is mainly squid, one of which can be seen on the deck near the punctured stomach. If the stomach of a whale-bone whale is punctured a ton of krill may gush out.

of sea-water swallowed during feeding; in the whale-bone whales this must be considerable. Fish deal with this excess of salt from the sea-water by having special salt-secreting cells on their gills, but whales lack these and how they manage to survive with no fresh water to drink and all this excess of salt to get rid of is still something of a mystery. It is thought that they must pass a great deal of urine (their kidneys are large) and get rid of the salt in this way. But of course a large amount of water is lost at the same time, so that it is rather surprising that the whale does not become dehydrated. Against this is the fact that whales do not lose water by sweating as we do, and lose very little vapour when they breathe out, because the air they breathe is so damp as to be almost saturated.

One of the questions that has troubled men for years is why the deep-diving whales never suffer from the 'bends'. This disease, also known as caisson sickness, is a great hazard to all human divers, and in the old days helmet divers were often crippled or even killed by it. The cause of it is as follows. When a diver breathes compressed air, the nitrogen in this air dissolves in his blood, and if he is then brought up to the surface too quickly this dissolved nitrogen comes out of solution and forms bubbles of gas in his blood, in the same way as bubbles form in a bottle of beer when the pressure is suddenly lessened by taking off the cap. Divers, such as the Polynesian pearl-divers, who have no breathing apparatus and simply hold their breath (as whales do) for as long as possible, are less inclined to this disease, for the air in their lungs is not at such great pressures, but they can still get the bends – which they call 'tarawana' – if they perform too many dives in too short a time.

Whales, and particularly sperm whales, dive far deeper than man and for longer periods throughout their entire lives, presumably without ill effects. It is now known that sperm whales can dive or 'sound' to a depth of 500 fathoms, for drowned sperm whales have been found entangled in deep-sea cables at that depth. Captains of whaling ships have always claimed that they could reach such depths because harpooned whales would pull out that length of line as they dived, but for a long time scientists would not believe that this was possible. The giant squid on which the sperm whale feeds, lives at great depths, but the plankton is found in the first five fathoms of the sea, and so most whales do not dive so deeply. Dolphins can apparently reach 400 fathoms. Sperm whales and beaked whales hold the record for staying under water the longest. They can hold their breath for 90 and 120 minutes respectively; human beings can hold theirs for one minute, although experienced divers can with practice manage up to two and a half minutes.

If one observes a whale sounding it can be seen that before it dives it will take a number of quick breaths, about one every ten seconds, whereas normally it will breathe only once every two or three minutes. These breaths can be seen, because each time it breathes out it 'blows', that is to say, it sends a plume of condensing water vapour (not water as many people think) and mucus shooting up into the air. Whalers claim that for every quick breath it takes before its dive a whale will stay under for a minute. When it resurfaces after a dive, it produces an enormous explosive blow, up to twenty-five feet high, and then swims near the surface for some time, breathing at very short intervals – in fact, it pants.

The whale's respiratory and blood systems are adapted to make possible these long deep dives with no ill effects. All these adaptations are not fully understood but I will try to describe what is known as simply as possible.

The lungs of cetaceans are not particularly large for their body size, but they are much more efficient than those of land animals. When we breathe we do not usually fill our lungs more than half full, and when we breathe out a large amount of this air is left behind in the lungs. You can test this for yourself, for after you have breathed out normally you will still be able to force more air out of your lungs. Even when you have finished this experiment there will still be about one litre of air left inside. Cetaceans, on the other hand, fill their lungs to capacity with every breath, and, moreover, change eighty to ninety per cent of

A rorqual inhales. The double blow hole, looking very like a pair of human nostrils, is set at the top of the head so that the minimum amount of body shows as the whale takes breath.

the air each time. Therefore, when cetaceans breathe, far more oxygen will diffuse into the blood than it does in human beings. This is then transported to the body tissues, for example the muscles, where it is needed for the processes which provide energy. In all muscles there is a pigment called myoglobin, which is akin to the haemoglobin of the blood but darker in colour, and which joins readily with oxygen and, in fact, stores oxygen until needed for metabolism. Cetaceans have up to eight times as much myoglobin as land animals, which means that whale meat is very dark in colour, but also, and more important, it means that the whale can store more oxygen, thus enabling it to hold its breath for longer.

This, however, does not explain why it does not suffer from the bends. Many explanations have been offered for this. At one time it

The 'blow' of a right whale. The blow of many whales is characteristic and is one of the few ways a whaler can distinguish his prey. Thus the right whale has a double blow, whereas the rorquals have a single vertical spout, and the sperm whale has a single spout inclined at an angle of about forty-five degrees. The blow consists of water vapour which condenses as it cools on escaping under pressure; it also contains oily mucus from the respiratory tract of the whale. The blow is said to corrode the skin of man.

was believed that whale blood contained special bacteria which would remove nitrogen. It has also been thought that the oily mucus in the air passages leading to the blow holes absorbs the nitrogen. Now the most likely explanation seems to be that the whales get the maximum oxygen into their system by 'panting' before diving and then take a minimum of air down with them, so that there is very little nitrogen under pressure in their lungs to dissolve into the blood. It is even possible that the air is isolated within the tubes of the lungs and thus prevented from passing into the air sacs, from where it would be absorbed into the blood.

Recently an experiment has been performed with a dolphin which was trained to swim to depths of 2,400 feet and there push a button which caused an underwater camera to take a picture. What is more, the dolphin was trained to breathe out into a special apparatus when it surfaced. This told us two things: when the dolphin was at depth its rib-cage was completely collapsed, and that during its dive it had used a very small amount of oxygen – just enough to keep its heart going.

The ribs of cetaceans are arranged rather differently from our own, and are presumably more mobile, and their lungs contain a large amount of elastic tissues.

The blood systems of cetaceans are also different from those of most land animals, in that they have in various parts of the body, particularly on either side of the spine and below the brain, a branching and twisting network of tiny blood vessels which are known as the *retia mirabilia,* or marvellous networks. These can obviously take up vast quantities of blood and it has been suggested by Professor Slijper that they may do this if one part of the body, the thorax for example, is under great pressure.

In addition, when a cetacean, or any other animal come to that, dives, its pulse rate immediately drops, in dolphins from 110 beats per minute down to 50, in ourselves from 70 to 35, and in seals from 120 to 10. Although the heart beat slows up, the arterial blood pressure does not fall, proving that some blood vessels are being constricted. The blood is being cut off from the muscles and the intestines, and is only supplying the vital organs, the heart and the brain, which cannot function without the oxygen brought by the blood.

The brains of whales and dolphins are large and comparable to man's, both in size per pound of body weight, and in the enormous folding of the cerebral hemispheres – the conscious part of the brain. They are intelligent animals; some people would say very intelligent, even intellectual animals. But their senses are very different from our own.

To begin with, the toothed whales at least have no sense of smell, and although the dolphin is said to have a good sense of taste it seems doubtful if the large whales have, judging by the miscellaneous collection of oddments found in their stomachs, including wood, feathers, paper, and even, in one instance, a bouquet of flowers.

The cetaceans sense of sight is not as well developed as our own. In some of the river dolphins it is practically non-existent. Indeed for deep-diving whales it must be of limited use, since below 215 fathoms the sea is pitch black, and even in the clearest shallow seas the maximum visibility is only 200 feet. Dolphins and porpoises have the best sight in the cetacean world and, in fact, hunt partially by sight. These eyes, which of course are adapted for seeing underwater, should

A female bottle-nosed dolphin gives birth to her young, which emerges tail first. Behind can be seen the female companion who will help during the birth and act as nanny to the baby. It is often the grandmother who plays this part.

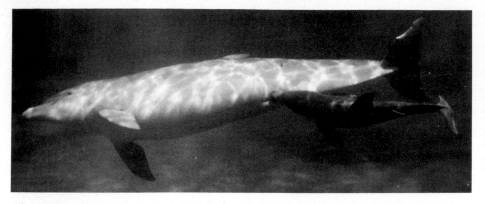

Above A young bottle-nosed dolphin is suckled by its mother. The milk is squirted into the baby's mouth from two teats on either side of the vent, which can be seen in the photograph.
Right For the first few months of its life the young dolphin will swim beside its mother. In the photograph, the umbilical scar is still visible on the baby.

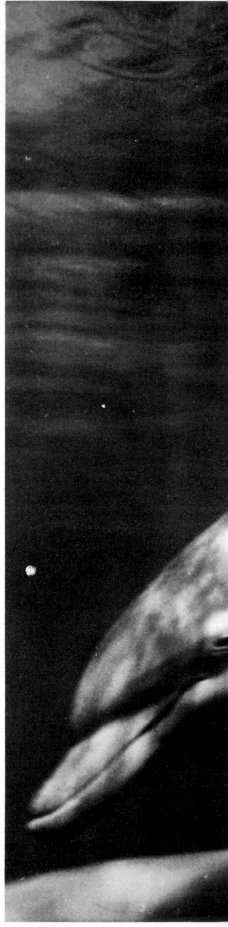

in theory be very short-sighted in air, and yet the dolphin can obviously see well, for he will leap through a hoop, or take fish from his trainer's hand up to twenty feet above water. It has been suggested that this is because the eye is lengthened by muscles outside the eye; this would be unique in mammals. The outer layer of the back of the eye, the sclerotic layer, is greatly thickened in cetaceans, probably to withstand pressure. The position of the eyes in the heads of the great whales, particularly in the sperm whale, is such that they can neither see directly in front or behind them, a fact which was made use of by the whalers when 'creeping' up to a whale, in rowing boats.

The sense of touch seems to be well developed. The skin is thin and easily damaged but dolphins like to rub themselves against smooth surfaces or be scrubbed with a brush in dolphinariums.

However, the cetacean's most important sense is its sense of hearing. It is difficult for man with his inadequate hearing to realize what is involved. It is like trying to describe perfect vision to someone who has been almost totally blind from birth. For the cetacean hearing does what our combined senses of hearing, sight, touch and taste do for us, because using his sense of hearing only, a dolphin can distinguish between two types or sizes of fish at the other end of its tank. It does this by a system of echo-location, very similar to our SONAR or ASDIC systems, where ultrasonic noises are given out and the time taken for the echo to bounce back tells us how far away any obstacle is. This is similar to the system that bats use for finding their way about, although, in this case, the medium is air.

The dolphin's sonar, however, is far more efficient than man's, which is one of the reasons why the dolphin has been studied so extensively recently (see Chapter Seven). Some of the noises thrown out are low enough to be heard by the human ear if underwater hydrophones are used, and sound rather like rusty hinges. In addition to these door-creaking noises the dolphin also whistles, moans, squeaks and barks; these noises are his way of communicating with other dolphins and, incredibly, he can carry out a 'conversation' at the same time as using his sonar system.

Yet if you look for these marvellous ears on a whale or a dolphin, all you will find, if you search carefully, is a tiny slit behind the eye of the animal, and in many species the ear-hole has become solid. The

Mother and baby take a look at what
is happening above the surface of the
water. Dolphins are born with all
their senses fully operative.

functional part of the ear is housed in a very hard lump of bone called the 'bulla' which is rather loosely connected to the rest of the skull, and therefore was often brought back by whalers as a souvenir.

In toothed whales the sounds may be made by the release of air through a series of valves and diverticula in the passage leading to the blow hole, or they may be made in the larynx. The bulbous forehead of toothed whales – the melon – is thought to have something to do with the sending out and receiving of sounds. The melon may also be very sensitive to touch and register changes in water currents, for it has a very good nerve supply.

Although all aquatic mammals have bodies which are adapted to some extent to their watery lives, most of them must come on to dry land to breed; but not the whale. Whales are so completely adapted that if they are grounded they will quickly die, for not only will they overheat but their skeletons cannot support their bulk when not held up by the buoyant water and once they are stranded their own enormous weight will crush their internal organs. So a whale must court, mate, give birth and suckle its young in the water.

The courting and mating of whales and dolphins is performed with great charm and joie de vivre. The male singles out a female and follows her around for days, posturing in front of her and stroking her with his body and flippers. As she becomes excited they will swim and leap and dive together and finally couple, belly to belly, sometimes leaping into the air as they do so, so that the mating of the large whales is a most spectacular sight.

Most infant whales are born ten or eleven months later, though sperm whales take a little longer. Very rarely twins are born, but usually, as in humans, there is only one infant. It emerges tail-first and is pushed up to the surface for its first breath, by its mother or another female, who has become the mother's inseparable companion during her pregnancy, and now acts as midwife and nursemaid to the young animal.

For the first three months the infant swims just behind the dorsal fin of the mother, where, because of the interaction of their pressure fields, it gets a boost forward.

Suckling presents certain difficulties under water, since it can only be done for very short periods between journeys to the surface to take breath. The teats which lie on either side of the genital slit are normally covered by folds of skin (as is the male's penis) for the sake of streamlining, but one is everted when the baby comes to suckle, and the muscles around the teat squirt milk into the baby's mouth. This is necessary since cetaceans do not have mobile lips and thus cannot easily suck, and it also speeds up the process of feeding. Cetaceans' milk is extremely concentrated and has a very high fat content; forty to fifty per cent, compared with two per cent in human milk and four per cent in cows' milk. It looks like condensed milk and tastes, or so we are told, like a mixture of fish, liver, milk of magnesia and oil.

Fed on this rich diet the infant grows at a prodigious rate. A blue whale is heavier at birth than an average sized elephant, and grows at a rate of one and three-quarter inches per day. At about four and a half years of age it will be sexually mature, and at thirteen it will have reached its full length of about eighty feet, a size that dwarfs even the largest dinosaurs.

THE WHALE-HUNTER AND HUNTED

Fish, I love you and respect you very much; but I will kill you dead before the day ends.
Hemingway (The Old Man and the Sea)

Man's attitude to the animals he hunts is strangely inconsistent. At the same time as we deeply deplore the present-day slaughter of whales we cannot help admiring the ingenuity and courage of the early whalers.

To primitive men a whale represents tons of fresh meat, oil for eating or lighting, bones for building, and sinews for rope and string. Presumably the earliest men lacked the skill or the implements to hunt whales and simply took advantage of stranded ones or of corpses washed up on their beaches. The earliest record we have of man's association with whales comes from Røddøy in northern Norway, and was scratched on rock by a Stone Age man some time about 2,200 B.C. The drawing shows a seal and two porpoises with a man in a boat; it is not clear whether he is hunting them, but it seems possible. Another wall drawing from Norway shows what is undoubtedly a whale hunt, for the animal is beset by men in canoes, and near the whale's tail there appear to be men upset from their boat into the water. It seems incredible that men in open canoes, armed only with implements of flint or bone, would have the temerity to attack anything as large as a whale, or that their primitive weapons could penetrate the thick coat of blubber and kill the animal. Actually we can make a good guess at how it was done by studying the methods used by the Eskimos, who, until the Europeans reached them, lived a pre-Iron Age life. The only raw materials that they had were the bone, ivory and skin of animals, with a small amount of driftwood, and, very occasionally, small pieces of meteoric iron.

We have a description, written in 1874, of Eskimo whale hunts. The hunts were conducted from open canoes with crews of eight picked men. The boat was paddled, although there was a sail with a mast which also served as harpoon and lance by having a harpoon head or a large knife lashed to it. The harpoon head was made of ivory, tipped with slate or iron, and attached to it was a line and a number of buoys of inflated seal-skin. These buoys would make it difficult for the whale to sound, and, consequently, it would tire more quickly. Each time the whale rose out of the water to blow, more harpoons with buoys would be fixed in it. When finally the whale was exhausted the man who first harpooned the whale was given the job of cutting a hole in its side, large enough to admit the lance which was used to administer the coup de grâce. The whale was then towed to shore, cut up and shared out. The Eskimos are of necessity a thrifty people, and every part of the animal was utilized. The favourite parts for eating were the flukes, lips, fins and flippers, the poorer meat being fed to the dogs. The oil was used for lighting and heating and to barter with the tribes of the interior – the 'reindeer men'. A seal-skin full of oil was equal in value to one reindeer. The bone was of great use for the making of implements, and every man in the crews of the whaleboats got a share of this in addition to his share of meat. Even the entrails were pickled and used as a relish and as an anti-scorbutic.

In those days these whales (presumably grey whales) were also hunted by the Red Indians from Vancouver and Queen Charlotte's Islands. They too employed floats on their harpoon lines, but used sharp pieces of mussel or abalone shell to tip their harpoons. Evidently these Indians delighted in the hunt and were exceedingly fierce and daring. Any Indian who had killed a whale was honoured by a cut across the

A method of catching whales supposed to have been used by the American Indian. (Facsimile from de Bry.) This dangerous and probably impossible method of killing whales by stopping up their blow holes and thereby suffocating them is also said to have been employed by the natives of Madagascar.

A female narwhal dragged on to the ice in Greenland. Only the male narwhal has the long, twisted tusk which was often made into a cane or walking stick and which probably gave rise to the unicorn legend.

nose, which was a mark of the greatest esteem.

Modern-day Eskimos still occasionally catch a whale and they, like the people of Eastern Siberia, are exempt from international restrictions, being permitted to kill the protected species such as the grey whale and the Greenland right whale.

North of Vancouver and Queen Charlotte's Island lies a string of islands, stretching between Alaska and the eastern coast of Russia, called the Aleutian Islands. Very few of them are inhabited nowadays, but, before the Europeans reached them two centuries ago and decimated the population, a primitive people called the Aleuts lived there. They had perfected an ingenious, if horrible, method of catching whales. They paddled out to them in fragile kayaks made of skin, planted a poison-tipped spear, and fled homewards as quickly as they could. The poison they used was fermented from the roots of aconite. The whale died in two or three days and if it was washed up on shore it was claimed by the man whose mark was on the spear.

This hunting was recorded in 1872 by Pinart, a French ethnologist, but he reports that the poison used was fat from the corpses of rich men! Whether the Aleuts were trying to preserve the secret of their aconite poison, or whether this was an alternative method of killing the whale by blood-poisoning, I do not know. Whale poisoning by aconite was also recorded in 1774 in the Kurile Islands north of Japan.

The Japanese themselves had an even more extraordinary means of catching whales, which called for great skill and daring. They caught their whales in nets. Before 1600 some whales had been caught by the Japanese simply with harpoons and lances, but at the beginning of the seventeenth century they invented this successful net method. The idea was soon copied all over Japan. Because large numbers of men were involved and the nets were costly, the enterprise was subsidized by a rich man. The whaling prospered and some of these men became very rich, owning a whale-processing concern as big as a village and employing large numbers of men. Their riches enabled them to entertain the men of culture of their day. Thus there are a number of writings and beautiful Oriental prints of this industry.

A whale was caught, in a net, in the following manner. A watch was kept on the hills overlooking the sea and when a whale was sighted, six net boats and many hunting boats set out with the sailors rowing in a standing position and the catchers, ready with harpoons, in the stern. The net boats worked in pairs and, taking up a position in front of the whale, they rowed away from each other, casting the nets between them. Then the men in the hunting boats would start to shout and yell and beat the sides of their boats and, approaching the whale from three sides, drive it into the nets. The whale would become entangled in the nets and halt – long enough to be harpooned by the hunters. The hunting boats would then be dragged by the whale and each time it surfaced it would be lanced by the hunters. At last, exhausted by the weight of the boats and the loss of blood, it would come to a stop and be killed – a sight 'terrible enough to make one break into a cold sweat', wrote Yosei Oyamada in 1829.

Amazingly, right, humpback and fin whales were caught in this way, but not grey whales, since they broke the nets. As fin and humpback whales sink when they die, a further hair-raising feat had to be accomplished by the net whalers before the whale was quite dead. One of the catchers had to jump on to the dying whale and with a long knife he would cut a hole through the snout, meanwhile rising and sinking below the surface of the water with the dying whale. Another catcher would dive into the sea and swim to the whale with a rope which he would thread through the hole in the snout before returning to his boat. The catchers wore their hair especially long so that they could be pulled by it out of the water and back into their boats. While all this was going on, the best swimmers would dive under the whale with ropes, so that the animal was suspended between the boats in a sort of hammock. Then, with final stabs from the lances, the whale went into its death flurry and was hauled ashore to the cries of 'May its soul rest in peace!' The final haul was done by shore-based capstans, and an army of flensers and meat carriers stripped the carcass.

The meat was salted and the bones boiled to extract oil. The residue was sold as fertilizer for the rice-fields. Special watchmen were employed to prevent the workmen and local peasantry from stealing the meat. Net whaling continued in Japan until 1909.

It seems incredible that an animal as strong as a whale could be checked by flimsy nets, and that all the whales did not destroy them and escape the whalers as the grey whales did.

Dolphins are also checked and confused by coarse nets, although they can perceive and avoid nets with a mesh less than ten inches square (in fact, they will leap in and out of fine nets in order to steal fish). It is thought that the beams of sound emitted by the dolphins' sonar will pass right through the coarse mesh and therefore the animal does not perceive the net as a continuous structure.

If fin whales have a sonar system this might be the explanation for the success of the Japanese net whalers, but it seems doubtful that they have. The structure of fin whales' skulls and voice-boxes is different from that of the toothed whales, all of whom use sonar. Also they do not have the same need for echo-location as do the toothed whales,

'Fishing off Zanzibar'. This picture clearly shows the whaling boat with four men at the oars and the boat steerer in the stern with the large steering oar. The possibility of a whale allowing a man to sit on its snout and hammer a spike in seems as unlikely as the feasibility of the suffocation method of a previous picture. The artists were presumably not whale men.

64

'Cutting in', 1574. The flensers are aided by a tune on the bagpipes. The artist had obviously never seen a whale, as is evidenced by the strange, square teeth, the tubular blow holes and the row of very human-like breasts.

since they do not usually hunt fish or squid, nor do they dive as deep and so probably are able to use their sight. Dr Purves of the British Museum thinks that whale-bone whales probably find the krill they feed on by their sense of smell (for the krill does have a smell) and therefore do not need sonar for hunting. He is of the opinion that this is the reason for their having double blow-holes, two nostrils being necessary for locating the direction of smells. The toothed whales have no organs of smell, and therefore have evolved a single blow-hole for efficiency in breathing. Perhaps the humpbacks, right whales and fin whales were simply *confused* by the unfamiliar sensation of being caught up in nets.

Certainly the confusion of whales and dolphins has been used by man in order to catch them at other times and in other ways. In Canada at the mouth of the St Lawrence, the belugas, which have come upstream at high tide in search of fish, have their way to the open sea blocked by rods which are placed upright in the water and which vibrate in the outgoing tide. Although the rods are ten feet apart and the belugas could easily escape between them, they do not attempt to do so, and are caught because the vibrations confuse them. This method was originally used by the North American Indians but the palisade has recently been rebuilt in order to capture belugas for aquaria. Similar traps were used in Denmark for catching the common porpoise.

Whales can also be attracted, repelled, or confused by noises. The earliest example of this is Arion attracting the dolphins with his shrill song; and the latest is the use of ultra-sonic noises by modern whalers to put whales to flight. This is done because the faster a whale swims the more frequently it must come to the surface for air, thereby presenting itself as a target for the harpooner.

65

Above The body of the killer whale illustrates the beautiful streamlining which is present to some extent in all whales and helps them attain surprisingly high speeds. The sharp, shark-like dorsal fin of the killer shows well here.

Above right A sperm whale on the flensing deck. The intestines of whales are usually dumped overboard, but the liver and various glands are often retained and vitamins and hormones are extracted.

Right The skull of the killer whale shows the massive and powerful lower jaw and the 'terrible array of teeth'. 'There is no remedy against an attack by a killer whale', wrote one diving expert, 'except reincarnation.' But, in fact, the killer seems to bear man nothing but goodwill.

The most curious use of sound is that employed by the people of the Gilbert Islands, where there is a special 'porpoise caller' or 'dreamer' who goes into a trance and then with a high whine like that of a puppy calls 'our friends from the West'. The unfortunate 'friends' swim inshore and ground themselves, only to be slaughtered. This was witnessed by Sir Arthur Grimble, who was administrator of the islands, and recorded in his book *A Pattern of Islands*.

Many different peoples drive dolphins or whales ashore by beating the sides of their boats, or slapping the water with hands or sticks, or beating stones together underneath the surface of the water. This behaviour has been observed in places as far apart as the Solomon Islands in the Pacific, and the Faroe, Orkney and Shetland Islands. The people of the Faroes still hunt the pilot whale or 'caa'ing whale', as they call it, in this way. The animals are driven towards the shore by men in boats, shouting and banging oil-drums and making as much noise as possible. Once the leading whale is beached the rest will follow, and the catch can be considerable, for pilot whales swim in large schools.

When toothed whales get into shallow water they seem to get confused or to panic. Probably their sonar will not function efficiently in shallow water, for the beams of sound seem to be directed horizontally and upwards only, and a gently sloping shore may not be perceived. In addition, mud or sand does not send back very clear echoes, and the surf will further confuse their sonar signals. This probably accounts for the natural strandings, which in the middle Ages were taken as portents of evil, and which sometimes involve

'Catching a whale off Deptford Pier', 1850. Occasionally fin whales do get in too close to land with disastrous results. In 1967 a fin whale was imprisoned by the falling tide in a land-locked pool in Newfoundland. The effect this had on the whale and the local populace is recorded in Farley Mowat's book, *A Whale for the Killing*.

really large numbers. Two hundred false killer whales *(Pseudorca crassidens)* were stranded in 1935 in South Africa; sixty-seven pilot whales in 1955 in the Orkneys; and seventeen killer whales in 1955 in New Zealand. The people of La Paz in the Gulf of California seem to have been particularly cursed by this phenomenon: on the 3rd of February, 1954, twenty-four sperm whales ran aground there, only to be followed on the 27th by another thirty-four! They say that the stench from the dead whales was unbearable.

A possible explanation for these apparent mass-suicides, other than panic, is that once the leading animal is grounded it calls and the rest follow in order to help it. But, whatever the reason, the instinct to swim towards shore in these circumstances seems to be extremely strong. The animals will even resist attempts by men to lure them back into deep water. Anthony Alpers tells of a valiant attempt by friends of his, throughout a whole night and part of the next day, to rescue a group of stranded dolphins. Every time an animal was pushed out into deep water it immediately swam inshore again. In the end they only managed to save two.

I can find no accounts of fin whales being stranded in large numbers, which could be circumstantial evidence for their not possessing a sonar

The capture of a whale at Broadstairs, 1850. Here again the artist had little knowledge of whales: the position of the eye is incorrect and it is impossible to identify the species of the whale from the picture. However, at least the scales, fins and fangs of mediaeval drawings have disappeared.

Far left A sperm whale is 'worked up' on the flensing deck of a factory ship. The dark, almost black colour of the flesh can be seen here. This is due to the large amount of myoglobin in the muscles, which is necessary to deep-diving animals. It also makes the flesh unpalatable.

Left The ribs of a fin whale are separated. They will be sawn up by a steam saw and dropped down hatches to the factory deck below to have the oil boiled out of them. Even the modern whaler's life is not without hazard. Without his spiked boots he might well fall on the flensing decks, thick with oil and blood, and fall into the boiling try-pots below with the bones and blubber.

Below An early fin whale foetus. In its early stages this foetus is much like any other mammalian foetus. The glove beside it gives an idea of its size. At birth it would have been about twenty feet long.

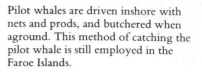

system, or could simply be because they do not hunt in shallow waters. But men have come across these whales marooned in tiny pools of Antarctic water, their way to the open sea cut off by the advancing ice. These poor animals perished by drowning, as their escape holes to the air froze up.

Men and whales do not only meet at either end of a harpoon. There are instances of men and cetaceans cooperating in the hunting of fish. This was first recorded by Pliny in his *Natural History,* where he states that the people of Nîmes in France used dolphins to help them catch mullet. At a regular season these fish left the marshes and swam by a narrow channel into the sea. It was impossible for the fishermen to spread nets there and so, when the fish started to run, the entire populace would come to the beach and call for the dolphins, which would appear in answer to this shouting and drive the fish back towards the shallows. Caught between the dolphins and the men with nets, not many mullet would have escaped.

More recently, in 1856, the aborigines have been observed hunting fish in the same way. A particular tribe, living appropriately at Amity Point on the Pacific coast of Queensland, had a very friendly relationship with a school of dolphins which, if fish were sighted, they called by slapping the water with their spears. The dolphins would then drive the fish shorewards as the latter tried to escape from the spearsmen.

The people of Mauritius fish in exactly the same way to this day. Here the dolphins are called by slapping the water with the hand or a piece of wood. Soon dolphins appear on the horizon and swim towards land. As they approach, the Mauritians go out into the water with nets and traps, and again the mullet are caught in the middle.

Both men and dolphins profit and it is difficult to decide whether the dolphin is helping man or merely using man to help itself. Perhaps this slapping of the water sounds like a leaping fish to the dolphins. Certainly if you want to teach a captive dolphin to feed on dead fish (which are easier to supply than live fish) you must slap the water with the fish and then place the fish under the water.

The river dolphins have long been made use of by fishermen in just the same way. Thus the bouto of South America and the Irrawaddy river dolphin *(Orcaella brevirostris)* of India and the Chinese river dolphin *(Lipotes vexillifer)* all take part in commensal fishing. In fact, fishing villages in India and Yunnan adopt dolphins for this reason, and it has been reported that at the end of the nineteenth century legal suits were often brought in the local Indian courts because the plaintiff's dolphin had been filling his neighbour's nets instead of his own!

The most extraordinary story of cooperation between cetaceans and man concerns the killer whale. It comes from Twofold Bay, Australia, which is on the Pacific coast, south of Sydney, and lies on the route taken by whales on their journey from the summer feeding grounds of the Antarctic, north, to their breeding grounds nearer the Tropics. There was a shore whaling station here between 1866 and 1928 and from this a family of early settlers hunted the migrating humpbacks and minke whales in open boats. In the month of July, when the whale-bone whales arrived, so did a pack of killer whales, which stationed itself at the mouth of the bay. When the killers sighted their quarry they would start 'lob-tailing' or 'flop-tailing', that is to say, beating their tails on the surface of the water – a noisy practice. This may have been a signal to alert the other killers in the pack, but the whalers took it to be a signal for themselves, and quickly launched their boats. The whale was approached on one side by the boats, and, on the other, by the killers who kept harrying the whale and driving it towards the shore, and this meant that a whale which would normally take up to twelve hours or more to chase and kill could be despatched by the whalers in an hour.

Once the whale was harpooned the killers played their most important part in this grizzly partnership. Four of them would station themselves under the head of the whale, preventing it from sounding, while the others swam on either side of it, throwing themselves one after another on top of it and covering its blow-hole. The exhausted whale was then killed, and the men attached an anchor to the harpoon line and left it to the killers for a couple of days. This, however, was not entirely altruistic for these whales sink when they die and do not rise again until inflated by gases of putrefaction. The sinking of the

Left New Bedford whalers on the
North West coast, 1843. Harpooning
a whale from a rowing boat was
always a risky job, and the cry 'A
dead whale or a stove boat!' seems to
indicate that the one was as probable
as the other.

Below left 'Stove Boat', a whaleman's
water colour which, like the
preceding illustration, is now in the
Whaling Museum, New Bedford.
Sperm whales were said to crush
boats or men in their jaws.

Below An abandoned Antarctic
whaling station at Prince Olaf
Harbour, South Georgia. As the
waters round the shore stations were
fished out men turned to pelagic
whaling.

whale did not incommode the killers, of course, and they took the tongue and the lips as their share.

There is something very unattractive about the role the killers played in this action. One feels an aversion to their siding with man against one of their own kind, but perhaps this is sentimentality or anthropomorphism. Certainly the men of Twofold Bay did not feel this way. They were fond of their pack of killers, and, when the last of them, Old Tom, died, he was not rendered down for oil; instead his skeleton was carefully cleaned and mounted, and put on show at the nearby town of Eden.

Left A sperm whale stranded at Gairdner River, Australia. The square head and narrow lower jaw with its peg-like teeth can be well seen. The upper jaw, although containing rudimentary teeth, has none that show through the gum, but merely sockets into which the teeth of the lower jaw fit.

Above Grounded pilot whales. All toothed whales get stranded from time to time, especially on gently shelving beaches which seem to confuse their 'sonar'.

Left A loaded harpoon gun in the bows of a catcher. Note the harpoon line and the telescopic sight for the accurate aiming that is essential in order to hit the small strip of back which is all that shows above the water.

Below A shore whaling station on the eastern Atlantic seaboard, 1969. One whale has already been hauled up the slipway ready for flensing, and more dead whales await their turn. Shore whaling stations like this one have operated all over the world for hundreds of years. As the number of whales decreased as a result of the greater efficiency of the pelagic whalers more and more of these had to close down.

Left A female grey whale tries to cut the rope attached to her harpooned baby. The barnacles on her skin (showing well on the flukes) make her back a good cutting implement. Mother whales will not forsake their injured babies, and in some species of whale, notably the humpback, the male will try to defend his injured mate. These facts were made use of by the whalers, who would harpoon the baby or the female first.

HEYDAY OF WHALING

'There she blows! There! There! There! She blows! She blows!'
'Where-away?'
'On the lee-beam, about two miles off! A school of them.'
Herman Melville
(Moby Dick)

In the last chapter I wrote of many strange methods of catching whales, but the most usual method was simply to spot the whales from the ship or shore station and then chase them in small fast boats. The opening quotation from *Moby Dick* conjures up the exciting moment when the whales were first sighted by the look-out at his station on the cross-trees of the whaling vessel. The whales were harpooned and lanced from the boat, and towed back to the shore or the mother ship to be cut up and rendered down for oil.

This method was used by the large European and American whaling industries, which prospered at different times from the ninth century until the twentieth century, and is still used in a more mechanized and streamlined form by the great whaling countries of today, Japan and the USSR.

In A.D. 890 King Alfred the Great, writing down the verbal account of one Ottar of Norway, tells us that whaling was taking place off northern Norway even in those days. In fact the name 'whale' comes from the Anglo-Saxon word *hwael*. It is very similar to the German *wal* and the Norwegian *hval* and is related to our modern word *wheel*. It refers to the turning movement that whales appear to describe as they breach and then sound, rising for air and then diving again in one smooth semi-circular action.

The Norwegians have continued to hunt whales right up until today. Theirs is a mountainous country, with little arable land but with an enormous sea-board, so that the flesh of whales has augmented their other supplies of protein foods. Indeed, as mentioned in Chapter Two, Norwegians settled in Iceland in order to avail themselves of the large numbers of whales in her waters.

As the Norsemen invaded other European countries they passed on their skill, and it was perhaps thus that the Basques learned the art of whaling. In any case, by the eleventh century the villages bordering the Bay of Biscay, particularly the towns of Biarritz and San

Sebastian, which are today fashionable seaside resorts, became thriving whaling communities. Many of the towns along the northern coast of Spain have a whale in their coat of arms to commemorate this. It was the Biscayan right whale which they hunted, for it did not swim very fast, and it floated conveniently when dead. The Biscayan right whale *(Eubalaena glacialis)* can be distinguished from the Greenland right whale *(Balaena mysticetus)* by the characteristic lump on the tip of its upper jaw known as the 'bonnet'. Now it is a rare animal, seen occasionally either singly or in small groups, but in those days it was found in groups of up to a hundred all over the North Atlantic, and particularly in the Bay of Biscay where it returned to breed in the autumn after spending the summer further north.

This industry continued until the sixteenth century and in fact prospered and spread to other parts of Spain, and to Portugal, France and England, and even to Newfoundland. There was a great need for whale products, particularly the oil, which was used for lighting. In addition there was much demand for whale-bone, not only for stiffening women's clothes but for many articles for which nowadays steel, elastic and plastics are used.

At first only whales near the coast were harpooned, so that the carcass was hauled ashore for processing, but as the whales became scarcer the whalers had to voyage further afield. It was impossible to

In the nineteenth century the sperm whaling industry reached its peak and sailing vessels set off from port to sail right round the world on whaling trips lasting as long as four or five years. The precious oil, seen barrelled in the foreground of the picture, was used in candles and oil lamps, and to lubricate the increasing number of machines and steam engines.

process a whale satisfactorily in a small boat and so the blubber was stripped off and brought ashore and the rest of the carcass abandoned to the gulls and the sharks.

In England the whale was declared a royal fish and the King proclaimed an honorary harpooner. The head of any whale caught in coastal waters or stranded on shore automatically belonged to him, and the tail to the Queen. The hunting of the Biscayan right whale was continued in the Hebridean 'fishery' until the twentieth century.

The hunting of the Greenland right whale started later (apart from those caught by the medieval Icelanders). In the fifteenth century Englishmen and Dutchmen sailed north-eastwards, trying to find a North-East Passage to the Orient, the southern route being dominated by the Spaniards and Portuguese. They did not find a route, of course, and therefore never got their oriental spices, but instead they found vast numbers of whales around Spitzbergen, Jan Mayan Island and Novaya Zemlya. So began the misnamed Greenland fishery (Spitzbergen was named East Greenland by the whalers).

The hunting of whales in the Arctic prospered and many nationalities took part in it. Eventually it spread westwards as far as Baffin Bay, and by the eighteenth century the Americans joined in the hunting, on both their east and west coasts.

Greenland right whales continued to be hunted until the start of this century. The last ship set off from Dundee in 1912 but came back empty-handed. There was not a single whale to be found!

As the number of Greenland right whales declined the Americans started to hunt humpback whales and then later sperm whales. The sperm whale industry thrived in New England in the eighteenth and nineteenth centuries. From Nantucket and New Bedford, from Martha's Vineyard and Cape Cod, a fleet of up to 735 ships sailed off on cruises which took them right round the world and lasted for several years. On one of these ships sailed Herman Melville, and from his adventures aboard whale ships came the novel, *Moby Dick*.

This marvellous book which can be read as an allegory, as a tragedy of Shakespearean dimensions, or simply as an exciting adventure story, tells of the gaunt, obsessed Captain Ahab. With a peg leg carved from the jaw-bone of a whale, he seeks vengeance on the albino sperm whale, Moby Dick, which on a previous hunt had bitten off his leg.

The novel also tells us a great deal about whaling in the 'sperm fishery', but it is all slightly romanticized. The reality was considerably more brutal and degrading. Melville himself, having been at sea since he was seventeen, signed on board the whaler *Acuchnet* in 1842, but deserted her in the Marquesas. After some anxious weeks living with the natives of the Island (they were reputed to be cannibals), he was taken off by another whaler, the *Lucy Ann*. The *Lucy Ann* was barely sea-worthy; the rigging was rotten, the crew either drunk or dying of scurvy or venereal diseases, the timbers crumbling (the cook often broke off splinters from the bitts and beams for kindling wood), the food was condemned navy stores purchased cheaply in Sydney, and the ship swarmed with rats and cockroaches. There follows his account of the last-mentioned horror from his novel *Omoo*.

'These creatures (the cockroaches) never had such a free and easy time as they did in her crazy old hull; every chink and cranny swarmed with them; they did not live among you, but you among

them. So true was this that the business of eating and drinking was better done in the dark than in the light of day.

Concerning the cockroaches, there was an extraordinary phenomenon, for which none of us could ever account. Every night they had a jubilee. The first symptom was an unusual clustering and humming among the swarms lining the beams overhead, and the inside of the sleeping places. This was succeeded by a prodigious coming and going on the part of those living out of sight. Presently they all came forth; the larger sort racing over the chests and planks; winged monsters darting to and fro in the air; and the small fry buzzing in heaps almost in a state of fusion.

On the first alarm all who were able darted on deck; while some of the sick who were too feeble lay perfectly quiet – the distracted vermin running over them at pleasure. The performance lasted some 10 minutes, during which no hive ever hummed louder . . . . Nor must I forget the rats; they did not forget me.'

Not surprisingly, when the *Lucy Ann* reached Tahiti, most of the crew, including Melville, mutinied, and were taken ashore to the local gaol, the Calabooza Beretanee.

This does not seem to have been an unusual pattern. Whaling vessels often returned with an almost entirely changed crew, new crews being picked up wherever the ship came to port, and the islands in the Pacific had a drifting population of dissatisfied whale men.

Whale men from Australia and New Zealand had the worst reputations. The first convict settlement had only been founded recently (in 1788) and the land was only very thinly populated with

'Destruction of the larboard boat of the *Ann Alexander* by a sperm whale in the South Pacific' from the Peabody Museum of Salem. In 1851 a bull sperm whale stove two of the *Ann Alexander's* boats, and when the captain retaliated, sank the mother boat as well. The crew were picked up after two days in the remaining boats. Not so lucky were the crew of the *Essex* which was stove by a sperm whale in 1820. The crew of one of her boats were finally rescued ninety days later. They had stayed alive by drawing lots, and then shooting and eating the loser.

'The whale in his flurry', 1874. The boats had to get as close as possible to the whale in order that the harpooner might hit him with the heavy harpoon. Ideally it would be 'wood to blackskin'. When the whale's heart or lungs were hit he would spout red blood and it was said by the heartless whalers that his 'chimney was on fire'. As the whale died he would 'go into his flurry' and then finally roll over on to his back, 'turning fin out'.

Aborigines and Maoris, whom the whale men had no compunction in killing. Now and again the natives would strike back, and in 1809, the entire crew of a whaler was killed and eaten by Maoris.

Thus whaling stations were set up at the Bay of Islands, and the Derwent Estuary in Tasmania, and many more were set up along the south, east, and west coasts of Australia to catch the whales migrating northwards in the autumn. The whaling towns became notorious for drunkenness and vice, for the whale men on the whole were a disreputable lot. They were a polyglot society of tattooed South Sea Islanders, Maoris, Red Indians and Azores or Cape Verde Islanders picked up on the long sperm whale voyages, mutineers, and other American and European outcasts, many of whom were escaping from justice.

A list of Melville's shipmates aboard the *Lucy Ann* paints the picture. There was Long Jim, Black Dan, Flash Jack, Jingling Joe, M'Gee,

Bembo, Wymontoo and Beauty, as well as Melville's particular friend, Doctor Long Ghost, the only other educated man aboard.

Of course there were respectable and well-ordered whaling ships, and the owners and even the captains of these became rich men. The crew were paid a 'lay', that is to say a percentage of the profits when the whaler returned and the oil was sold. Ishmael, the hero of *Moby Dick*, was offered a three hundredth lay. Thus if the sailor could last out the whole voyage he had a lump sum to return to – provided that they made a good catch, or that he was not drowned, or that the ship was not lost in a gale or 'stove' by a whale.

Sperm whales were notorious for attacking vessels. Any whale could damage a boat with a stroke of its tail but sperm whales would also ram boats with their great square heads or even crush men and boats in their jaws. It seems to have been the lone males who became ship sinkers and they earned formidable reputations and were known by name, such as, Mocha Dick, Timor Tim, Don Miguel and New Zealand Jack (who should not be confused of course with Pelorus Jack or Opononi Jack).

Life on board a whaler was undoubtedly rugged, even if the captain was not brutal or drunken. The quarters were cramped and the food after the first few weeks at sea consisted of rock-hard bread or ship's biscuits and 'salt horse' (meat pickled in brine which gave off a fetid smell when lifted from the barrel). A few live animals, such as pigs or chickens, were taken on board, but these were for the captain's table. Sometimes the cook would enliven the diet with duff, which was a pudding made by boiling flower and water dough, or the sailors would fry their biscuits in the rendering whale oil in the 'try pots'. Occasionally they would eat whale meat; but sperm whale flesh is almost black in colour and considered by most people inedible. Of course there was the daily ration of spirits.

In addition to these hardships there were long periods of inactivity and boredom, suddenly changing to frantic action and danger when whales were sighted. The sailors combated boredom in various ways. They made, out of whale-bone or whales' teeth, small articles such as walking sticks, pastry crimpers, snuff-boxes, buttons, thimbles and chessmen, with rather crude pictures scratched or carved on them. This work is known as scrimshaw or scrimshander. They would also dance and play and sing, and drink if they could get hold of the liquor. They would play practical jokes on each other too, such as hauling someone into the rigging feet first. If another whale ship was sighted they would probably hold a 'gam' – a party where news and mail were exchanged and a good deal of alcohol was generally consumed.

However, with the call of 'There she blows!' everyone sprang into action. The small open whaling boats which were always kept ready were quickly launched with cries of 'A dead whale or a stove boat!' and, using sail or oars, the whale was approached.

Every whaling boat launched from a whaler had its captain or 'boat header' and its harpooner or 'boat steerer' as well as a crew of four others to man the oars. As they approached the whale the boat-header steered in the stern with a long steering oar and issued whispered instructions to the crew, all of whom had to row as silently as possible, with their backs to their quarry. Imagine the terror of approaching, backwards, a possibly belligerent animal bigger than a

The whale monument by Knut Sten at Sandefjord commemorates the dangers and courage of open boat whaling.

house! The harpooner, who was the most respected man on board the whaler after the captain and the mates, would stand in the bows, his knee braced against the clumsy cleat, a notched board in the bows, and throw the harpoon. This needed great strength, for the harpoon was heavy and had to be flung a distance of up to thirty feet. The harpoon was about eleven feet in length with a barbed metal tip so sharp that Queequeg, the Polynesian harpooner in *Moby Dick*, used his to shave with! Attached to the other end of the harpoon was a thin Manila line. This had to be enormously long (about 200 fathoms), since sperm whales are notoriously deep divers. It was spirally coiled in tubs in the stern of the boat. From there it ran around the loggerhead (a post in the stern of the boat), down the length of the boat resting on the handles of the men's oars, and through a groove in the prow where it looped and was coiled a number of times. It was finally attached to the two harpoons kept in readiness on a crotch in the prow. The other end of the line was not attached to the boat, as this could have meant the boat's following the whale below in a deep dive; it was left free to be attached to the tub of line in the next boat if necessary.

Once the harpooner had struck the whale with his harpoon he would immediately try to fix the second one in case the first did not hold. He was not always successful in this since the whale dived as soon as it was hit. Then the second harpoon would thrash about becoming a danger not only to the whale but also to the crew of the boat.

As the whale dived the line would run from the tubs at an amazing speed, hissing over the wrists of the oarsmen, and smoking as it ran around the loggerhead and through the groove in the prow, so that the men had to douse it constantly with sea-water to keep it cool. The arrangement of the line was a very important matter and a harpooner would take a whole morning carefully stowing his line; the slightest tangle or kink in the rope could take off a man's arm or leg as the whale ran, and men were quite frequently tossed overboard by the running line. As Melville says: 'The six men composing the crew pull into the jaws of death with a halter round each neck.'

Once the whale began to tire the boat-header would loop the line once or twice around the loggerhead and hold the animal, and the boat would be dragged at a crazy speed on a 'Nantucket sleigh ride', so that sometimes the sailors did not know if they were travelling on water or through the air, and all they could do was clutch the sides of the boat and pray. Eventually the whale would stop and lie gasping for air, exhausted at the surface of the water.

At this point the harpooner and boat-header changed places in the boat – 'a staggering business truly in that rocking commotion' – for it was the boat-header's privilege to kill the whale. This was done with darts – large spear-like implements with long wooden handles. The whale was stabbed until a lucky blow hit lungs or heart, and the whale died in a flurry of white foam and red gore spouting from its blow hole.

This process, from the harpooning to the death, could take a whole day or longer.

If a school of whales were encountered as many as possible were harpooned quickly, using the old Indian method of attaching floats or 'druggs' made of wood to the harpoons and coming back to finish off the wounded whales at leisure. Once these were killed they were marked with a 'waif', a long poled flag, and it was then the job of the

'Cutting the blanket', an old photograph of the *Charles W. Morgan* which first sailed from New Bedford in 1841. The cutting in stage, like a painter's cradle, can be seen and also the long-handled blubber-spades and the tackle attached to the rigging for raising the great pieces of blubber. In the nineteenth century whales were flensed like this, at the side of the boat, the blubber was rendered on board and the carcass left to the sharks.

mother ship to go round collecting the carcasses.

Open boat whaling as described above is still carried out to this day in the Azores.

Sperm whales float when dead and so they could be secured by chains at head and tail to the mother ship, and a platform with a hand-rail was lowered over the whale for the 'cutting-in' or flensing. The men stood on this platform and with long-handled, sharp-edged blubber spades cut off the head and proceeded to peel the blubber from the animal by making an incision in the blubber above the side fin and a semicircular cut below it. Into this hole was inserted a large hook, attached to a heavy tackle hung from the main mast. From this a rope ran to the windlass and, by dint of many men's heaving away at the windlass, the blubber was peeled from the whale in a great spiral strip known as a 'blanket piece', the whale turning over and over in the water as this was done. The harpooner had the honour of fixing the blubber hook into the blubber, and in order to do this he had to descend on to the whale as it lay half-submerged. Often he would remain on the revolving creature to free the blubber from the underlying tissues with his spade. In this amazing circus feat he was in danger not only of drowning but also of being crushed between the whale and the ship, or of losing a foot to the sharks which, attracted by the blood, would be attacking the carcass.

Once the blubber and the head were off, the carcass was left to the sharks. This was enormously wasteful as about thirty per cent of the whale's oil is found in its bones.

The blanket pieces of blubber were now cut into smaller horse pieces, and then chopped up finely by the mincers with great double-handed knives and fed into the try-pots.

The try-works – the apparatus used for rendering down the blubber – was situated on deck between the foremast and mainmast. It was a solid structure of bricks and mortar and contained two iron furnaces which heated the two great round iron try-pots, each of which was large enough to accommodate a supine, curled-up man. Beneath the furnaces were placed large pans of water to prevent the wooden deck from catching alight. The furnaces at first burned wood, but, once the rendering had got under way, the shrivelled remains of the blubber,

Whale-bone scrapers from Hull and Whitby, 1813. In those days whale-bone was as precious as the oil. It was used for making such articles as umbrellas, whips, portmanteaus, chairs and sofas, and carriage springs, as well as being used to stiffen ladies' dresses and corsets. Now, with the guts, it is the only part of the whale thrown overboard as useless.

which were known as 'fritters', were fed into the fire.

The smoke and smell must have been appalling. According to Melville, 'it has an unspeakable, wild, Hindoo odour about it, such as may lurk in the vicinity of funeral pyres. It smells like the left wing of the Day of Judgment; it is an argument for the pit.'

In fact by the time the oil was stored in barrels and stowed in the hold, the crew were well covered with grease and soot, and a great cleaning operation took place, when the try-pots, the decks, the sails and the men's clothes were all washed and scoured.

The head of the whale got special treatment. If a whale-bone whale had been caught the skull and upper jaw were cut out and lifted on to the deck. The baleen was separated and the gums scraped off the bases of the plates, which were tied in bundles and then stored. The head of the sperm whale was precious too, but for a different reason. The lower jaw, which is made of very hard ivory-like bone and which holds the teeth, was retained and used for making canes and other articles, and the teeth were kept for scrimshaw work or for trading with the natives of the islands they visited. But the most precious parts of the head were the case and the junk, which together make up the great square head of this whale. These heads were usually too heavy to be lifted on to the deck and had to be dealt with half in and half out of the water at the ship's side. The lower half of the head, known as the junk, is an immense honeycomb of tough elastic tissue and contains a large quantity of oil, but it is the upper part or case which holds the precious spermaceti wax, which was used for high quality candles and for lubricating machinery. The case was emptied by means of a bucket on the end of a long pole.

If a whaler was lucky, ambergris might be found in the belly of the whale. Ambergris is a grey crumbly substance, sometimes found in the lower intestines of the sperm whale or sometimes expelled by the whale to float on the sea. It has the property of fixing perfume and for this reason fetched a high price. In 1912 a Norwegian whaler sold a lump of ambergris for £23,000, thereby saving the owner company from bankruptcy. Ambergris is still used to some extent in perfumery but can now be replaced by synthetic substitutes. In earlier times it was thought to be an aphrodisiac.

Ambergris was always supposed to be found only in sick whales. Melville had the following to say on the subject.

'By some, ambergris is supposed to be the cause, by others the effect of the dyspepsia in the whale. How to cure such a dyspepsia it were hard to say, unless by administering three or four boat-loads of Brandreth's pills, and then running out of harm's way, as labourers do in blasting rocks.'

However, when a record 926 pound lump of ambergris, larger in size than a man, was taken from a whale in 1954 by the whaler *Southern Harvester,* it was reported that the whale was extremely healthy and well-fed. So the reason for the occasional presence of ambergris remains a mystery.

However brutal and brutalizing life aboard a sperm-whaler was it at least had the advantage of warm weather, for, although the hunting took them into both the Atlantic and Pacific Oceans, they remained in temperate or tropical waters.

The American, Dutch and English whale men who hunted the grey and right whales in the Arctic, during the late eighteenth and early nineteenth centuries, had intense cold, iced-up rigging, and the very real danger of being trapped or crushed by the shifting ice to add to their discomforts. However they seem to have been a resourceful and jolly lot. We have accounts of ships being unloaded on the ice and hove-to so that the keel could be repaired, and also of cricket matches on the ice floes between teams from rival whale ships. Here is an account, written by a sailor in 1863, of the crew of a sister ship which had been split by the ice:

> '. . . I found her crew as happy as larks when I visited them on the ice. On my first visit the ship was perfectly upright, and some of the men were aloft, acting monkey just to amuse themselves. On my second appearance the ship had changed her position, and her masts and yards were now on the ice. This, however, made no difference to the crew; there were several of them on each yardarm of the main top-gallant yard having a see-saw . . . .'

The whales were hunted, killed and processed in very much the same way as they were in the sperm fishery, except that here there was natural refrigeration and the blubber was not rendered down, but simply cut up small, carefully stripped of skin and underlying muscle, and taken home in barrels to have the oil extracted at the shore station.

By the end of the nineteenth century the whaling industry had gone into decline, due partly to the shortage of right and grey whales, and partly to the fact that the use of whale oil for lighting had been superseded, initially by petroleum and later by gas and electricity.

However there was only short respite for the whales whose destiny it was to be turned into soap, margarine, and pet food during the twentieth century. Three unrelated happenings soon enabled man to hunt the whales with greater efficiency and vigour. The first of these was the opening up of the Antarctic. In 1774 Cook returned from his exploration of Antarctic waters and reported that they were swarming

A sixty foot sperm whale awaits flensing at a shore station. Sperm whale fishing flourished during the nineteenth century, but as the oil lamp was superseded this 'fishery' declined. Recently the numbers of sperm whales killed per annum has increased as a result of the virtual extinction of the large fin whales at the hands of the modern whaler.

with seals and whales. The second was James Watt's invention of the steam-engine which led to the building of powered boats which could chase the faster fin whales. The third was the invention of the harpoon gun. In 1840 a harpoon gun named the Greener gun was developed on Tyneside, and in the 1860s a Norwegian, Svend Foyn, perfected his own gun and invented a harpoon with an explosive head which incorporated a time fuse, and so exploded inside the whale. This 'most fortunate, religious and good old man, respected and beloved by all who met him' patented his invention on Christmas Eve and wrote in his diary, 'I thank Thee, O Lord. Thou alone hast done all.' Whoever was responsible, these inventions, coupled with man's greed, will probably lead to the extinction of at least one species of whale in the near future, if not to the extinction of all the great whales.

At the beginning of this century many nations set up whaling stations in the Antarctic, on South Georgia, in the South Shetlands, the Sandwich Islands and the South Orkneys, and, using fast boats and the new harpoon gun and exploding harpoon, they began to hunt the fast fin whales which until then had remained unmolested. These whales, especially the giant blue whale, provide a large amount of oil but have the disadvantage that they sink when dead and have to be inflated with gas to keep them afloat.

As the shore stations of the Antarctic fished out the waters near to them a method of chasing the whales further afield had to be invented. Thus pelagic whaling came into existence. Large factory ships fitted with a ramp at the stern for hauling the whale aboard, and boilers for rendering oil, could now follow the whales wherever they went. Small fast boats were used to catch the whale which was processed while still at sea. It is the efficiency of these factory ships and their huge cost which have spelled doom for the whales, because once businessmen have sunk money into building and equipping a pelagic fleet they will operate that fleet in order to recoup their money even if it means killing the goose that lays the golden egg.

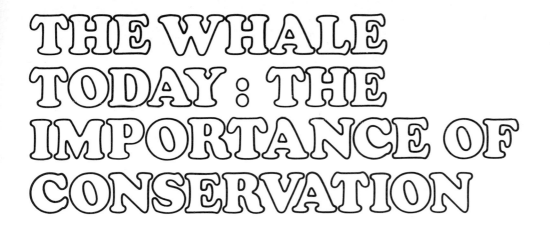

THE WHALE TODAY: THE IMPORTANCE OF CONSERVATION

The moot point is, whether Leviathan can long endure so wide a chase, and so remorseless a havoc; whether he must not at last be exterminated from the waters, and the last whale, like the last man, smoke his last pipe, and then himself evaporate in the final puff.
Herman Melville
(Moby Dick)

The modern factory ship is almost more than a floating factory, it is more like a small industrial town, being self-sufficient and giving employment to large numbers of men, to sailors, gunners, flensers, engineers, blacksmiths, carpenters, chemists and airplane pilots.

Each factory ship is accompanied by a number of small fast diesel-engined catcher boats, equipped with radio and sometimes with sonar to locate whales. In the bows of these boats are mounted the harpoon guns with the explosive-headed harpoons (although whalers have experimented with electric harpoons and curare-tipped harpoons in an effort to find a way of bringing faster and possibly more humane death). As whales have become scarcer each factory ship has employed more and more catchers. In addition there are usually a couple of 'buoy boats' to collect the whale corpses which have been marked by the catcher with a flag and a radar device, and inflated with gas so that they remain afloat. These transport them to the factory ship. There may also be a small fleet of boats to carry fuel oil to the factory ships and return to port with whale oil, and even a seaplane or helicopter to help in the search for whales, some factory ships having a special flight deck astern of the flensing deck. Japanese factory ships are also often accompanied by a refrigeration ship which stores the meat destined for human consumption.

In these factory ships the entire whale is processed, a far less wasteful business than the primitive flensing on board the nineteenth century whaler, as described in the previous chapter.

The blubber is stripped and rendered for oil, and the meat is either frozen for human or animal consumption, or cooked to provide meat extract which is used as a basis for soups and the meat-meal which can be used in animal foods. The skeleton is cut up by great steam-saws and turned into bone meal for fertilizers, after the large amount of oil in the bones has been extracted. Enzymes or hormones to be used for medical purposes are extracted from parts of some organs. Whale meat

Left Seven fin whales and one blue whale at the stern of a factory ship await hauling on deck. Flocks of cape pigeons, common birds of the Antarctic, feed on the offal thrown from the ship.

Below A rorqual on the slipway of a factory ship. As these whales sink when dead, their bodies have to be inflated with air. This shows particularly in the throat area where the grooves, like accordion-pleating, open up to give the mouth a baggy outline. When the whale is alive these grooves are tightly compressed so that the animal has an almost concave throat line, thus giving it an elegant streamlined shape and allowing it to swim at great speed. The rorquals are thought to expand these grooves while feeding in order to take in the maximum amount of krill.

has never been very popular in Europe for it has a unique taste. In England it was promoted by the Ministry of Food during World War Two but was so uninviting it became a national joke. Japan is now the only country that eats whale meat to any great extent, but even there the taste for it is said to be dying. Various parts of the whale are eaten by the Japanese, the most tender part coming from the tail at the root of the flukes, and this is eaten raw with soy sauce and grated ginger or horseradish.

One would think that European countries would have little use for whale products, but in 1971 (the latest year for which we have figures) 15,526 tons of whale meat and 13,228 tons of whale oil, largely sperm oil and spermaceti, were imported into Britain. What became of these vast quantities, which are calculated to account for between 2,300 and 2,800 whales? Most of the whale meat has been fed to our cats and dogs, largely in the form of tinned animal food, although some has gone into cattle and poultry feeds, and some has even ended up in our soups and gravies! The oil has been used for myriad purposes, but mainly in the beauty business – in our soaps, shampoos, creams and lipsticks. It is difficult to decide which is more ridiculous; to exterminate a species of animal in order to produce an artificial population explosion in another species or in order to decorate the skin of a third!

The conservation society known as 'Friends of the Earth' who are campaigning for the protection of whales have carried out a great deal of research into the use of whale products and in their view there is not a single substance produced from the whale for which there is no substitute. However, whaling is a 100 million pound sterling industry, with one whale being killed somewhere in the world every twelve minutes, and this industry is not going to be easily persuaded that its activities are unnecessary or undesirable. Indeed, in a world with a human population which is increasing at an immense speed it does not seem practicable to afford complete protection to whales when each one is such a potentially enormous source of food for humans. What is obviously needed is some sort of control which will prevent the over-fishing of whales so that they may still be around for our grandchildren and their grandchildren.

There has, in fact, been some control of international whaling since 1946. In that year the representatives of fourteen governments got together in Washington, and the International Whaling Commission (I.W.C.) was formed. This commission is still extant and so, just about, are the whales which it was set up to protect and conserve. Ever since 1946 the I.W.C. has tried to prevent the over-exploitation of whales, and since 1952 it has had scientific sub-committees studying the whale population so that sensible limits can be set for the numbers of whales killed each year and the whale stock may not be depleted to an extent that will make whaling impracticable in future years.

Unfortunately the work of the I.W.C. has been hindered in several ways. Firstly, not all nations belong to it; Peru, Portugal, Spain, Chile and Brazil do not. These nations are therefore not bound by any restrictions on catches that the I.W.C. may decide on. Furthermore, any nation which is a member can opt out of any ruling it does not like. For example, when it was decided in 1961 to ban the hunting of the humpback in part of the Antarctic, Russia, Japan and Norway, the only three nations with whaling fleets stationed in the Antarctic, all

Far left A rorqual is transferred to the factory ship of a pelagic fleet by an old catcher, now used to collect and transfer dead whales to the mother ship, thus leaving the newer catchers free to hunt and kill the whales.

Left The minke whale, the smallest of the rorquals, does not produce much oil but provides the best whale meat. It is reputed to taste like beef.

Below The tail of a sperm whale is securely stropped to the catcher. The enormous flukes, the up and down movements of which propel the whale with such force, are often cut off by superstitious whalers who are said to believe that if this is not done the dead whale will swim away. Dead sperm whales are often used as fenders by the catchers when refuelling from the mother factory ship, so that they will not smash each other to bits in rough weather.

opted out, making a nonsense of the resolution. Secondly, the I.W.C. has no teeth; that is to say, it cannot enforce its decisions at all efficiently. The captain and crew of a whaler taking a forbidden whale can be fined, but the company owning the boat still profits from its capture. Until 1972 the inspectors aboard the whalers who are responsible for the enforcement of the I.W.C.'s decisions have been of the same nationality as the whaler and therefore could not be said to be impartial.

The third factor hindering the conservation proposed by the I.W.C. is a decision they made themselves. In 1964 it was decided to protect all calves and nursing females and prohibit the killing of all right and grey whales. In addition they prescribed certain seasons for killing different species of whale and certain areas where whales could not be hunted, and set a definite quota of whales to be killed by each nation. It was in the calculation of this quota that the mistake was made of using a unit known as a Blue Whale Unit, or B.W.U. Instead of setting a specific limit to the catch of each kind of whale, a nation was allowed to take a definite number of Blue Whale Units, where a B.W.U. equalled one blue whale, two fin whales, two and a half humpbacks or six sei whales. This strange equation was worked out according to the amount of oil produced by each species, but as it was easier and more economical to find, chase and kill one whale rather than six, blue whales were taken in preference to any other species. Soon there were so few blue whales left that the catch had sunk almost to zero. In the 1964-65 season only twenty blue whales

A whale-bone whale on the flensing deck of an Antarctic Russian whaler. The baleen will probably be discarded, although riding crops are made of it to this day, but the lower jaw will be sawn up and the oil boiled out of it. Up to a third of the oil obtained from the whale is found in the bones, which are full of a fatty, yellow bone marrow.

were killed, and as the efficiency of modern whalers is such that any whale sighted can be caught, this presumably means that only twenty blue whales were actually seen.

The reports of the I.W.C. make depressing reading; the restrictions they agreed on seem always to have been too little and too late, and, in a sense, this body itself is a threat to the survival of the great whales, as its existence gives a false sense of security to the public. Thus, the Minister for Fisheries and Food in 1972, in squashing an attempt to ban the import of whale products into the United Kingdom announced: 'I believe that the Commission is the only body from which effective and successful protection for whale resources can flow.'

By 1972 the position was this: the blue and humpback whales were now protected (in addition to the grey and right whales) but their populations had been reduced to approximately one twentieth of their original numbers and they were so rarely seen that the whaling nations were not really making a sacrifice. There is some doubt whether these species will recover from this slaughter or whether they will become extinct. The fin and sei whale population had also been decimated so that there was some fear in scientific circles for the survival of the former species. The heat was now being turned, predictably, on the smaller minke whale and on the sperm whale. The I.W.C. was still struggling without success to get rid of the old Blue Whale Unit and to implement an international observer scheme whereby whaling ships would carry an inspector from another nation. Some countries, notably those of South America, still did not belong to the I.W.C., and in fact at least one I.W.C. country (Japan) was operating an extra whaling fleet under a Brazilian flag of convenience.

At this point public opinion began to take a hand, and at the United Nations Conference on the Human Environment at Stockholm in June 1972 a resolution was passed calling for a total moratorium on whaling; that is to say, demanding that no more whales should be killed for ten years. Needless to say it was impossible to get the whaling nations to agree to this. As the I.W.C. chairman put it: 'Stockholm was an expression of opinion: here we are in the real world.' However the resolution gave whalers a salutary shock, and at the next I.W.C. meeting the old B.W.U. was at last scrapped and replaced by specific quotas, an international observer scheme was agreed, and definite quotas were agreed for sperm and minke whales which had virtually not been protected before.

In addition the United States, under its Endangered Species Act, has banned the import of all whale products. There is some pressure in the United Kingdom to try and pass the same kind of law, for Britain, although paying lip service to the work of the I.W.C., and not a whaling nation herself any more, still imports large quantities of whale products, some from countries which are not members of I.W.C. and can therefore kill any whales that they can get hold of, regardless of age or species! However, since the Stockholm meeting the British Pet Food Manufacturers Association have announced that they have made their final order of whale meat, which is good news for the whale. How ironical it is that animal lovers are indirectly responsible for so much slaughter and so much suffering, for make no mistake about it, the killing of whales is often slow and undoubtedly cruel.

So, although the prospect for whales seems brighter now than it has

Above Although often kept in the same aquarium as dolphins, killer whales will attack and eat dolphins in the wild. Their diet also includes seals, penguins, fish, squid and even the large whales such as the grey and humpback whales. They are particularly fond of the lips and tongues of the great whales and will attack the harpooned whales towed by the catcher or buoy boats.

Above right The killer whale hunts its pray in a pack. The fact that it is a pack animal (as is the dog) probably explains why it is so easily tamed, and may also explain why a single killer in a dolphinarium will not attack the dolphins in the same tank.

Right The killer whale, like all cetaceans, is very graceful under water. Several killers kept together will perform a sort of underwater ballet.

Left A humpback whale surfaces. These animals do not really have a humped back, but got their name from their habit of showing a larger amount of back than other whales, when surfacing to blow. Although their shape and bumpy skin may appear grotesque when seen on the deck of a whaling vessel, underwater they are extremely graceful, using their long, wing-like flippers to turn. They are, in fact, the acrobats amongst the great whales, often leaping out of the water, and even turning somersaults in the air.

Right A pigmy sperm whale from the Miami Seaquarium. The thin dark skin is easily broken to show the white blubber below. Such injuries inflicted by the teeth of a whale's companions often look like blackboard drawings.

done for many years, nevertheless the questions remain. Have we reduced the population of some species of whales to such an extent that they cannot recover and must inevitably become extinct? And are the restrictions that we are now imposing on whalers sufficient to prevent the same thing happening to the species of whales which are still abundant? Or will we hunt each species to extinction, turning each time to a smaller species, finally exterminating the porpoises and dolphins? In fact, *can Leviathan endure so wide a chase?*

There are those who argue that the economic factor in whaling will save the whales; that nations will stop whaling as the whales get scarcer, for it will not pay them to catch them, and that the whale stocks will then recover. Certainly the number of whaling nations has decreased over the last twenty-five years and there are now only two nations with really big whaling fleets – the USSR and Japan, and it is rumoured that the USSR is finding it uneconomical to run their fleet and is going to cease whaling in the next three years. Whether the whale stocks will recover nobody really knows. The grey whale which has been protected since 1946 has increased in numbers, but no appreciable recovery of right whales seems to have taken place, and they have been protected for just as long. Nobody really knows how many of these whales there are left; certainly they are extremely rare but they are still extant. A group of breeding southern right whales (*Eubalaena australis*) has been discovered to visit the coast of Patagonia in the spring and autumn. These are being studied by scientists with a kind of eleventh hour, panicky intensity. They are protected by Argentina, but once out of her coastal waters are in constant danger from non-I.W.C. whalers. The Greenland right whale (*Balaena mysticetus*) is very rarely seen, and in 1963 when a Norwegian ship sighted this animal near Novaya Zemlya it was recorded in the log as 'a species that the crew had not previously seen'.

The blue whale population is not in very much better shape. Some

The fact that a dolphin can 'stand on
its tail' in the water and even 'walk' in
this position shows how powerful and
efficient are its tail muscles and
flukes.

Top White-sided dolphins at the
Oceanic Institute, Hawaii. Dolphins
spend much of their lives apparently
playing. They seem to leap out of the
water simply from joie de vivre.

Above A dolphin's tricks include
catching balls or rings, retrieving
objects from the water, swimming
and leaping in formation, jumping
through hoops and giving rides to
humans, as well as the somersaulting
shown in this photograph.

scientists think its numbers have fallen so low that it cannot possibly survive. This may also be true of the humpback and the fin whale. The trouble is that it is very difficult to work out whale statistics accurately. The information on sightings comes in the main from the crews of the whaling vessels who are bound to be biased, and, what is more, are likely to be in the seas where the greatest numbers of whales are to be found. The scientists of the I.W.C. are aiming at catches which will allow the whale population to recover and reach a level where whales may be caught each year without the overall population dropping. This is known as the 'maximum sustainable yield'.

Factors which have to be taken into account in working this out are the numbers of whales born and the frequency of birth for any species, the age of sexual maturity, and the social structure of the whale communities which will affect the ease with which the whales of any species will mate.

The blue whale, for instance, only produces one young every second or third year, and the sperm whale one every three years (its gestation period of sixteen months being longer than that of other whales). Neither of these young will reach sexual maturity for another four and a half to five years, and the estimated number of calves that a female can give birth to if she should manage to live to old age – a fairly unlikely event – would be ten for the fin whale, and only six for the sperm whale. This does not compare well even with the human reproductive rate. It will be seen that with this sort of breeding rate it will take the protected whale populations a long time to recover, if indeed it is possible for the various species to do so. The grey whale and the humpback whale have a faster reproductive rate, which probably explains why the grey whale population is rising, and augurs well for the survival of the humpback.

I hope that the humpback at least will survive for it seems to be the most charming and extrovert of the great whales. It is greatly given to lob-tailing and breaching or throwing itself into the air apparently out of sheer joi de vivre. It is also the most vocal of the large whales, producing lowing, grunting and yelping noises of enormous volume. It has even cut a disc, 'Songs of the Humpback Whale'!

Dr Payne who has studied these whale noises thinks that the high frequency notes are used for near communication while the low frequency notes will travel up to 700 miles under water.

The problems of studying the great whales are enormous. The problems of keeping them in captivity for study are insurmountable, for even a six-month-old whale will eat half a ton of fish a day. The only live whales which have been studied in any sort of detail at all are the grey whale and the killer and pilot whales.

The two latter are now being kept in dolphinariums in various countries and have proved their intelligence and friendliness towards man, although the unfortunate Robertson family who were sailing round the world had their yacht sunk by a pack of not so friendly killer whales in June 1972 off the Galapagos Islands. A young grey whale, Gigi, has also been kept in captivity in San Diego. It was she that ate the half ton of food a day. She has proved very docile, even cow-like, and quite different in character from her smaller cousins. This is also true of the right whales which have been studied off the Patagonian coast. They have been found to be docile, playful and friendly to man, lifting the scientists' boats gently six inches into the

Left Dolphins established a friendly relationship with their trainers and complete trust exists between them. The dolphin which is so graceful in water is helpless on land and completely at the mercy of man.

Above In addition to producing many ultrasonic sounds, dolphins can 'sing', making rather weird, shrill, squealing noises.

Above The depth of the blubber of a dead whale is measured. Some scientific studies are undertaken aboard factory ships, although this is often more perfunctory and hurried than the scientists would like as the main aim of the captain is to get the whales 'worked up' as quickly as possible.

Right An albino killer whale, aptly named Moby Dick, and his mother at Vancouver Island.

air with their great flukes and as gently lowering them back into the water.

This is the new image of the great whales; no longer would we agree with Job:

'When he raiseth up himself, the mighty are afraid: by reason of breakings they purify themselves.

The sword of him that layeth at him cannot hold: the spear, the dart, nor the habergeon.

He esteemeth iron as straw, and brass as rotten wood.

The arrow cannot make him flee: slingstones are turned with him into stubble.

Darts are counted as stubble: He laugheth at the shaking of a spear.

He maketh the deep to boil like a pot: he maketh the sea like a pot of ointment.

Upon earth there is not his like, who is made without fear.

He beholdeth all high things: he is a king over all the children of pride.'

Nowadays the fear has gone, indeed the crowds that gather around a beached whale seem to treat it more with contempt as they carve their initials and messages on its soft skin and stick a cigar in the blow-hole. However we are not all vandals, and those who care may save the whales. For it is possible that where the scientists have failed public opinion may persuade the whalers to spare the great whales for posterity.

THE PLAYFUL DOLPHIN

In the world of mammals there are two mountain peaks, one is Mount Homo Sapiens, and the other Mount Cetacea.
Teizo Ogawa

The opening quotation is how Professor Teizo Ogawa of the University of Tokyo describes the intelligence of whales and dolphins. At the moment there is a great deal of scientific controversy about the extent of the intelligence of this animal group. There are those who rate them with dogs, and others who rate them with chimpanzees, while yet others think them equally as intelligent as men. Some people, perhaps bewitched by the dolphins' charm, think that very soon they will be talking with us. Everyone, however, agrees that they are intelligent animals and have large, complicated brains. Since the opening of dolphinariums all over the world enormous opportunities have been offered to scientists for studying these animals and a great deal of research has been done, so that we now know far more about the dolphins, particularly the bottle-nosed dolphin, than we do about all the rest of the cetaceans put together. But the point that Professor Ogawa was making is that, living in so different a medium, the dolphins' intelligence is very different from ours, or that of a dog or a chimpanzee – so foreign to us that it is difficult for us to comprehend it, let alone measure or compare it.

In 1938 a studio-aquarium was set up at Marineland in Florida. It was originally intended as a film studio for shooting water and underwater sequences. A few dolphins were kept there and right from the start people would come to watch the dolphins being fed. Such is the charm of this animal. It was a short step from this to the dolphin shows, where the dolphins, and later the small whales, went through a routine of tricks to entertain the millions of people who came to watch them. Since then dolphinariums have been set up in a large number of places all over the world and the dolphin has been watched at work and at play, making love and giving birth, in sickness and in health.

Let us now consider the life story, from birth to maturity, of a

Dolphins can jump up to twenty feet above the surface of the water. Their jumping is so accurate and their eyesight so good that they can take an object in their teeth from out of their trainer's mouth without injuring him.

dolphin. The dolphin is born into a school with a definite hierarchy. There will be a leader of its little group, either male or female. In the wild, however, as Dr Norris reports from Hawaii, the really large groups have no leader. In the wild too there is very little fighting within a school, but in captivity there seems to be a more rigid pecking order, and more fighting to establish this. This is probably due to the confined space.

For the first few months the infant will swim beside its mother and be fed with her rich, oily milk. As it gets older it will make more and more sorties away from her side. Like most young animals it will spend a great deal of its time playing with the other young dolphins in the school. In fact it will spend a great deal of its life playing. Dolphins' food is abundant in the wild, as well as in captivity. They make no home or burrow and have few enemies – the shark, the killer whale and man being the only predatory animals large enough to attack them.

The dolphins show a superior intelligence in their play. They will leap or chase each other or play a sort of 'tag', and they will throw balls or rings, or anything that they can find, to the watching humans. As they initiate these games it could be said that they were teaching the humans tricks rather than the other way round. Even the hunting of a pack of killer whales is like a macabre nursery game as they circle their prey and take it in turns to make a murderous sortie into the centre of the circle. Dolphins are also great teases. For example, a dolphin placed pieces of fish near to a crevice where a grouper was hiding only to snatch them away as the fish was tempted out of its lair. They will also tease their trainers and the divers sent in to clean out their tanks – they will grab their implements or tip them over backwards in the water. But this is not really aggressive behaviour, for dolphins have powerful tails, hard snouts and mouthfuls of formidable teeth which could stun, disembowel or maul a man with the greatest of ease if they desired to do so. Happily for us the dolphins never do desire this.

The most extraordinary story of teasing comes from Marineland of the Pacific in California, and was reported in the Journal of Mammalogy by Brown and Norris. Two dolphins were trying to dislodge a moray eel from its crevice in order to play with it. After repeated unsuccessful attempts, one dolphin left the eel, killed a scorpion fish, which has sharp dorsal spines, and returned with it in its mouth. With its 'tool' it poked at the eel, apparently prodding it with the spines, and finally forced its unwilling playmate into the open.

The use of tools is unusual in the animal world – apart from man there are few animals that have learnt their function, but this trick of the dolphins shows something else. It shows not only an ability to reason out a problem but imagination too. There are other examples of this in literature concerning dolphins. For instance, a dolphin trying to obtain a piece of squid which had drifted under a bridge of rock hit upon the idea of fanning it out with its flukes, and two dolphins playing a game of 'pooh-sticks' with a pelican feather in the current from the water inlet, found that placing the feather in the eddies, rather than in the main current, gave them time to beat the feather and be in a position, further down the tank, ready to catch it in their mouths.

Thus dolphins' play is imaginative and inventive, and the best of the dolphin shows make use of this.

A dolphin can only be made to work by a system of rewards. If

Training must be done with a system of rewards, since dolphins do not take well to punishment. Killer whales are so intelligent and cooperative that they can be trained without food rewards.

punishments are used the dolphin will simply swim away and refuse to have anything more to do with the trainer. It may even refuse food and allow itself to die. If the dolphin becomes bored it will become destructive or morose. The tricks that seem to please the animals best are the creative ones. For example, at the Oceanic Institute of Hawaii a game was invented for a young female dolphin where she was only rewarded if she performed a new trick. She quickly grasped this idea and with obvious pleasure invented a number of new movements, such as spiral swimming, back somersaults, or swimming with her tail sticking out of the water! So, to our young dolphin, its eighteen months or so of training and its performances of circus tricks will be play rather than work. It will enjoy them.

A killer whale performing with a girl swimmer at Seaworld in San Diego objected to her cutting short his performance so much that, as

Two scientists from Marineland, Florida, listen to the noises made by a bottle-nosed dolphin over a hydrophone. Some noises that dolphins make are always accompanied by the release of air bubbles. The dolphin does not, however, breathe out under water, but rises to the surface to blow.

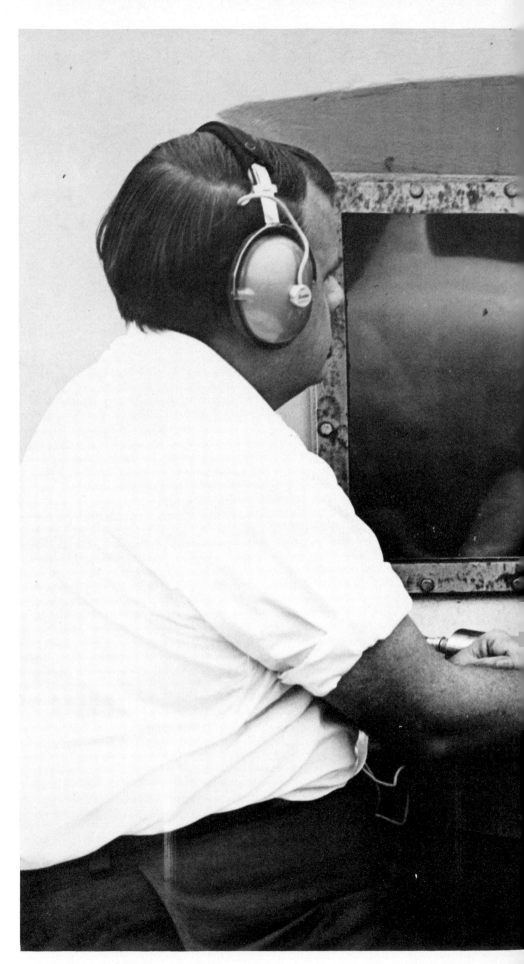

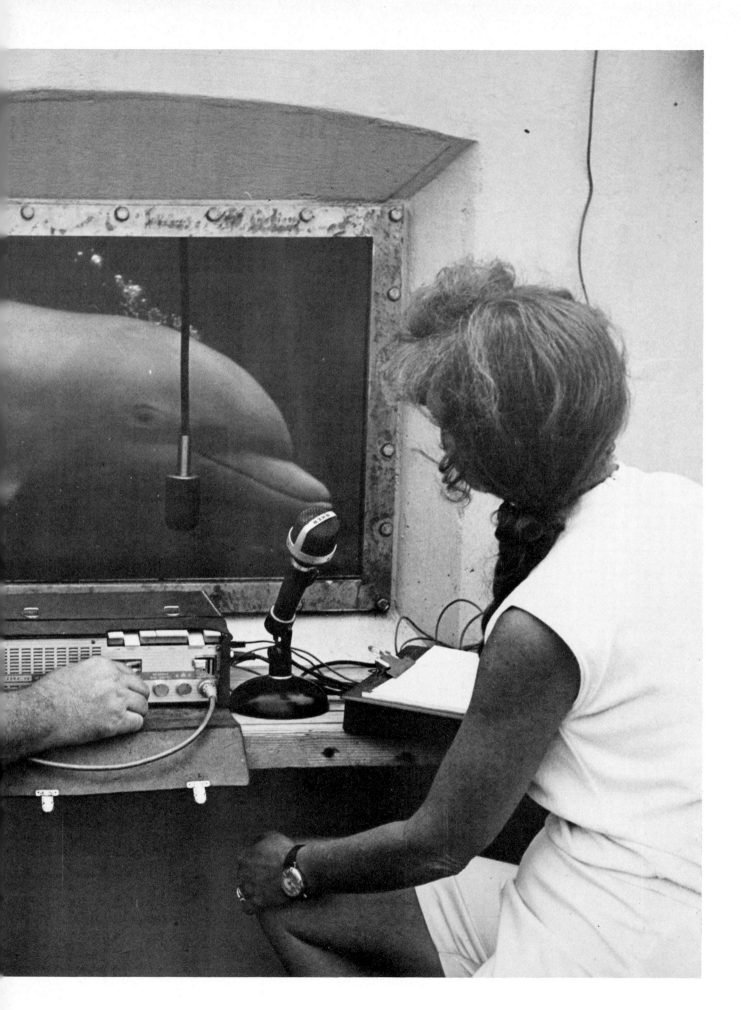

she made to leave, it grabbed her by the leg in its great mouth and
dragged her back, much to the horror of the spectators who expected
her to be eaten at any minute. However the skin of her leg was barely
broken – a gentle manoeuvre for a species that is purported to have
had the remains of as many as fourteen seals and thirteen dolphins
found at one time in its stomach.

If a dolphin that has been performing or been taking part in a
programme of experiments is rested, it will become morose and refuse
food, and must be given attention.

The world of our young dolphin is mainly one of sound. A great
part of its large brain (larger than that of a man) is concerned with the
transmission and reception of sound. Part of this brain is concerned
with the dolphin's sonar, the narrow beams of high frequency sound
which it emits through its forehead or melon. Some of these can be
heard by man as clicks and creaks, but most are ultrasonic. These
sounds are bounced back by solids in their path and received by the
dolphin through the lower jaw and ear. The low frequency sounds are
used for far objects, the high frequency for near. In this way a dolphin
finds its way around and hunts its food. It can distinguish between two
sizes of object that a man can only differentiate by using measuring
calipers, or, more usefully, between a fish and a water-filled plastic
'fish' of the same size.

Many experiments have been conducted to prove that the dolphins
'see' with their sonar. For example, when obstacles were put in the
dolphin's tank it faultlessly avoided them, even though blindfolded
with rubber suction cups, and when fish behind glass screens were
shown to a dolphin it could not find them, since the sound-waves
would not penetrate the glass.

In addition to the sonar the dolphins communicate with each other
by whistling sounds and they can even make shrieking noises in air
to communicate with man.

Dr Lilly, a neurologist who has done much work on the brains of
dolphins, claims that the dolphin mimics human speech 'in a very
high-pitched, Donald Duck, quacking-like way'. Now many species
of birds and even dogs and chimpanzees can be taught to mimic, but
they do not understand what they are saying or even (in most cases)
associate the phrase they are repeating with any meaning or action.
This is a far cry from dolphins actually talking to men, and yet Dr
Lilly claims that 'Within the next decade or two the human species
will establish communication with another species: non-human, alien,
possibly extra-territorial, more probably marine.' The press has
nicknamed him 'the man who will make fishes talk'! But if he should
prove to be right what a number of ethical problems this will raise!
Could our whaling fleets go on killing animals which could answer
back?

When Dr Lilly started his programme of research with dolphins he
was amazed to find how quickly they learnt and how willingly they
cooperated. There follows a story of a dolphin that cooperated to such
an extent that it was actually experimenting with Dr Lilly instead of
the other way round.

Dr Lilly was making a dolphin produce a whistle of given pitch,
duration and intensity, in order to gain a reward. Every time it
whistled it was possible to see the blow-hole twitch. After a few
minutes the dolphin became bored with this game and added a new

rule. It raised the pitch of each successive whistle. Soon it was out of Dr Lilly's acoustic range, and although he could not hear the notes, he could still see the twitching of the blow-hole. The rewards were stopped and the dolphin returned to whistles that Dr Lilly could hear. After that it never whistled out of the range of the human ear.

The inter-dolphin language seems to be very complicated. There are a large number of calls which humans are beginning to be able to identify, for example the babies' call on separation from their mothers, the S.O.S. call given by injured or sick dolphins, the 'I am bored' signal whistled by dolphins separated from dolphin or human companionship, the 'keep-together' signals of a school, the mating-call, and the 'hands-off' warning from one male to another. In addition there seem to be far more sophisticated communications. Each dolphin has its own signature whistle whereby other dolphins can identify it, and if a dolphin conversation is recorded on an oscillogram it can be seen that they do not whistle at the same time but take it in turns, as though listening to each other. Just how sophisticated the dolphin's communication is can be seen from the following examples.

A group of five dolphins swimming off the coast of California came across a complicated piece of sound equipment sunk across a channel by Dr John Dreher, who was studying the Pacific grey whale. As they approached the dolphins were emitting their routine sonar grating noises. At a distance of 400 yards they sensed the apparatus, bunched together and fell silent, as dolphins will in the presence of an unknown object which may prove dangerous. One dolphin then broke away from the group, came up to and inspected the hydrophones with its sonar and returned to the waiting group. A great deal of whistling took place. This was repeated three times and then the group quietly resumed their course and swam past the apparatus. Presumably they had decided that it was harmless. The second example concerns the dolphin's cousin, the killer whale. In 1957 the Norwegian Antarctic fishing fleet was bothered by a pack of killers a thousand strong, which were decimating their catches, and so the fleet summoned the aid of some catcher boats from the nearby whaling fleet. One killer was harpooned and injured, and within half-an-hour there were no killers in the vicinity of any of the gunboats, although there were plenty still wreaking havoc amongst the fishing vessels. The extraordinary thing about this story is that the catcher boats and the fishing vessels were identical – all converted World War Two corvettes. The only difference was that the catchers had harpoon guns mounted in their bows.

How can an animal without a complex descriptive language transmit the message that must have been passed by the wounded whale to its companions?

The fact that vocal instructions can be passed from one dolphin to another has been proved experimentally by Dr Bastian of the University of California. He separated two dolphins, Buzz and Doris, by a curtain so that they could not see each other but could still communicate by sound. Doris was then shown a signal that Buzz could not see, either a continuous light or an intermittently flashing one. If the steady light shone, Buzz had to press a lever on the right-hand side of his tank before Doris could press hers and be rewarded with a fish. If it was the flashing light then Buzz had to press the left-hand lever before Doris could receive her reward. Doris got her fish!

An Amazonian river dolphin or bouto is transported to Fort Worth Zoo. To prevent its own weight from suffocating it, the animal is carried on a litter lined with foam rubber. Throughout its journey the dolphin's faithful attendant keeps it damp so that it will not die of heat stroke.

A very large number of tape recordings have been made of dolphin noises and these have been analysed in an effort to identify some words or coding that would add up to a dolphin language – so far, without success. 'What we need,' said one researcher, 'is a Rosetta Stone for dolphins!' These recordings have been played back to other dolphins to try and learn the meanings by watching their reactions. This work had proved very difficult to interpret and has led scientists to wonder if their language has no words or symbols like ours. Dr Norris has suggested that by using their echo-location system dolphins can read the emotions of other dolphins in the air-spaces of their heads. This sounds incredible but can be compared with humans' understanding each others' meanings by their facial expressions, tone of voice, and by the pauses between words rather than simply by the meanings of the words themselves.

As our young dolphin grows up it will form attachments with other dolphins. These may be platonic love for another of the same sex, or a love-affair with the opposite sex that will lead to mating and the birth of a dolphin pup. Two male dolphins separated from each other at Marineland were reunited after some weeks. For two days these dolphins swam and leapt together, with obvious pleasure in each other's company, ignoring the females in the tank. This was not a sexual relationship, although dolphins do indulge in homo- as well as heterosexual activities, or even try to mate with large fish, or use inanimate objects for masturbation. They are not sexually inhibited as we are.

Examples of friendships between the sexes are numerous. Here is one of the most touching. In 1954 a small female common dolphin, Pauline, was caught with a hook and line and put into a holding tank. She was suffering from shock and, after trying unsuccessfully to rally her with shots of adrenalin her captors kept her afloat with a pair of water-wings fashioned out of four large empty jars, for, if a dolphin becomes unconscious and sinks, it will die from drowning. She would undoubtedly soon have died had not a large male dolphin been captured and put into the tank with her. Immediately, however, she showed signs of life and started to try to swim. The buoys were removed and with some support from the male she began to swim around the tank. The two became inseparable, but unfortunately the wound from the hook festered and two months later Pauline died. The male was inconsolable; he circled her body whistling, and after it was removed continued to circle the tank whistling most of the time. From this point on he refused food and after three days was found dead. On post-mortem examination he was found to have died from a perforated gastric ulcer. Perhaps it would have been more true to say that he had died of grief.

The maternal love of cetaceans is equally strong. Whalers in less enlightened times knew this and took advantage of it. If a calf was harpooned, the mother would come to its rescue and could also be killed. Mother dolphins have been known to push their dead babies to the surface and hold them there for days.

In addition our dolphin will have a strong corporate feeling for any member of its school – or, indeed, for any dolphin, even one of a different species. No sick dolphin suffers alone. Another dolphin will come to its aid, and push it to the surface and support it if its swimming is weak. According to Dr Lilly dolphins will do more than this.

One of his experimental dolphins (the same one that had played the vocal game with him) had been kept in a small tank for several days during cold weather before it was released into a larger tank with two other dolphins. The cold and the restraint had caused its back to stiffen up into an S-shape and it could not swim properly. It immediately started to give the dolphin S.O.S. call and the two other dolphins swam to it and lifted its head out of the water. A great deal of twittering and whistling took place between the three dolphins, and then the two healthy ones started to swim under their injured companion, raking the base of its tail with their dorsal fins. This friction caused the injured animal to bring his flukes downwards in a reflex jerk, thus forcing itself up in the water. They kept this up for several hours. This is further proof of the complexity of the dolphin's language.

In the sixteenth century a French zoologist, Pierre Belon, wrote of the dolphin: 'If he dominates and commands others, it is through his virtue and not through strength of arms.' This, it seems to me, pinpoints the difference between the most intelligent land animal and the most intelligent animal of the sea. It is sad, therefore, that man should now be considering the use of dolphins and small whales in warfare. To use them for peaceful purposes under the sea is another matter, and in fact they have been used for tool-carrying and the guiding of divers in several underwater experiments. But the United States Navy is at the moment using dolphins for purposes which are undisclosed, and a group of them have recently been brought back from Vietnam! The Russians also are running a research programme on the military use of dolphins. It seems likely that they are being trained for submarine detection and for the carrying and placing of mines.

It seems particularly ironical that the peaceful dolphin should be made a tool in man's urge to destroy his fellows, especially if it gets blown up itself in the process.

INDEX